INDUSTRIAL APPLICATION OF BIOTECHNOLOGY

INDUSTRIAL APPLICATION OF BIOTECHNOLOGY

I. A. KRYLOV

AND

G. E. ZAIKOV

EDITORS

Nova Science Publishers, Inc.

New York

LIBRARY OF CONGRESS CATALOGING-IN-PUBLICATION DATA

Industrial application of biotechnology / I.A. Krylov and G.E. Zaikov (editors).
p. ; cm.
Includes bibliographical references and index.
ISBN 1-60021-039-2
1. Biotechnology--Industrial applications.
[DNLM: 1. Biochemistry. 2. Biotechnology--methods. TP 248.3 I42 2006] I. Krylov, I. A. (Igor' Aleksandrovich), 1947- II. Zaikov, Gennadii Efremovich.
TP248.2I53 2006
660.6--dc22 2006003696

Published by Nova Science Publishers, Inc. ✣ New York

CONTENTS

PREFACE

Alcohol oxidase, glucose oxidase, and glucoamylase/glucose oxidase biosensors have been designed to analyze ethanol, glucose and starch concentrations, respectively. In chapter 1, the study of parameters of designed biosensors (limits of detection, time and accuracy of analysis, operation and storage stability) leads to a conclusion that they are practically applicable as analyzers for the systems of automatic control and monitoring at different steps of technological process in alcohol production.

Parietochloris incisa is a unicellular, fresh water green alga, capable of accumulating high amounts of the valuable very long chain polyunsaturated fatty acid (VLC-PUFA) arachidonic acid in triacylglycerols (TAG) in cytoplasmic oil-bodies. When growth is retarded by stress conditions, such as high light or nitrogen starvation, the biosynthesis of lipids is enhanced at the expense of protein or carbohydrate production. Thus, under nitrogen starvation, TAG accounted for over 30% of dry weight (DW) and AA content became as high as 60% of total fatty acids. To find out the cultivation conditions providing maximum yield of AA, the effects of light irradiance and N- availability on the DW and AA accumulation have been studied in chapter 2. From the standpoint of biomass accumulation, a light intensity of ca. 400 $\mu E\ m^{-2}\ s^{-1}$ was found to be optimal for growing *P. incisa* on complete medium. Lower light intensities (or higher cell density of inoculum) were found to result in high yield of AA when cultivated on nitrogen-free media. In the absence of nitrogen, algal cells were unable to cope with high light and suffered from photooxidative damage, whereas the nitrogen-sufficient culture survived under such illumination conditions probably due to accumulation of carotenoids. Nitrogen-deprived *P. incisa* cells displayed elevated sensitivity to light.

The identification of dehydroshikimic acid accumulated in fermentation broths was carried out with the use of different analytical methods as described in chapter 3. The capillary electrophoresis was used for determination of dehydroshikimic acid in fermentation broths and reaction mixtures of enzyme reaction. These experiments showed, that *B. subtilis* I-118/pDP4 strain had the best shikimate/dehydroshikimate ratio.

The technological scheme of production of the enzymes such as amylase, lipase and a complex of proteases from a cattle pancreas has been elaborated in conditions of its complex processing with isolation of ribonuclease. Chapter 4 has established, that isolation of enzymes is carried out according to the uniform technological scheme including stages of extraction, ultrafiltrate concentration and precipitation from water-salt and water-alcohol solutions. The optimum conditions of the given process realization have been chosen.

The technological unit for cultivation of pure alcohol yeast culture in aerobic conditions with guaranteed maintenance of aseptic conditions is offered in chapter 5. The main part of the unit is a bioreactor with inserted cartridge with tube type membranes made of nonporous polymeric material. The membranes of solubility are selective in relation to components of a gas mixture. In order to increase the economic characteristics of the unit, the block of preliminary enrichment of air with oxygen on the basis of use of liquid membranes is installed in its structure.

Research and development of novel types of reactors for the heterogeneous biocatalytical processes were carried out in chapter 6. The main directions for the reactor design were the significant intensification of mass transfer of substrate to immobilized enzyme and elimination of stagnant zones. The vortex reactors – *rotor-inertial bioreactor (RIB)* and *vortex-immersed reactor (VIR)* were developed for the diffusion-controlled biocatalytical processes. The lab-scale setups of these reactors were studied in dextrin hydrolysis that was limited by external and internal diffusion of high-molecular weight molecules of dextrin to glucoamylase immobilized on the carbon-containing inorganic supports. It was found, that under the optimal operation conditions, the process productivity and the observed activity of the heterogeneous biocatalysts were an average of 1.2-1.5 times higher in the vortex reactors than those in the fixed-bed reactor.

The kinetics of extraction of ribonuclease (RNA-ase) from bovine or other cattle pancreatic gland has been studied in chapter 7. This stage of the technological process of RNA-ase manufacture has been investigated and the mathematical model of the process has been developed. The process of RNA-ase extraction has been shown to conform to the first-order equation. The optimum conditions of extraction have been determined: the temperature of 5°C, pH 2.5 and 15% bovine pancreatic cells (by dried weight) providing 60% output from its content in bovine pancreatic cells.

Conditions of extraction of RNA and reduced forms of NADH from yeast *Sacharomyces cerevisiae* biomass were determined. Chapter 8 established, that the optimum conditions of RNA and reduced nicotinamide coenzyme extraction from bakery yeast biomass are the temperature 80 °C; pH 8.0, the time of extraction 20 minutes, and sodium sulphite concentration 0,5 mol/l, the RNA output being 70 %, and the reduced nicotinamide coenzymes output being 91.2 %.

The system of operative analysis of aerobic growth and biosynthesis in phase space of metabolic speeds is developed in chapter 9. Energy-releasing and energy-consuming metabolic blocks are represented in the equations of a constructional exchange with weighting coefficients proportional to generation and consumption of energy. Since stoichiometric equations are linear, qualitatively different areas in phase space of metabolic speeds are separated by linear boundaries. On this basis the graphically clear form of displaying the information on process of an aerobic biosynthesis on the display of the operator-technologist workplace is built.

In chapter 10, the selective system is offered to reveal the mutant plant cells characterized by the changed isoprenoid biosynthesis regulation at the level of HMG-CoA reductase. 25-Hydroxycholesterol, the effective inhibitor of HMG-CoA reductase, synthesized in Vitamin Institute RAS (Moscow), was used as a selective agent. To increase the frequency of genetic changes, high-effective chemical mutagen N-nitroso-N-methylurea (N-NMU) was used at the dose of 0,5 mM·h. These conditions of cell selection were approved by using *Dioscorea deltoidea* cell suspension (dedifferentiated cell culture) and several lines of morphogenic long-fibre flax callus culture. The repeated treatment of *Dioscorea deltoidea* cell strain PPI DM0,5 (characterized by furostanolic saponin superproductivity) was also carried out using the some dose (0,5 mM·h) of N-NMU. Irrespective of provided selective conditions, the frequency of genetic changes, which determined *Dioscorea deltoidea* cell resistance to 25-Hydroxycholesterol, was about 10^{-7} of plated alive cell number. During the period of active growth, we noticed colour variation (from light to greenish and yellowish) among Dioscorea cell clones. These selective conditions also very effectively allowed to reveal the mutant cells in control cell population.

The methods for immobilization of the urease on biosorbents obtained by surface activation of pyrogenic silicon dioxide – aerosil and microcrystallic cellulose using of an albuminous complex of a casein. The catalytically stable and active ferment preparations were obtained in chapter 11. The effect of pH and temperature on activity of both dissolved and immobilized ureases was investigated.

Characteristics of sewage waters forming under treatment of different types of raw materials were shown in chapter 12. The influence of synthetic surface active substances (SSAS) of different chemical nature on bacterial suspensions properties were investigate. Ecobiotechnological degreasing process allow to reduce the level of toxic contamination of sewage waters and provide optional elimination of greases from hair and skin tissues of sheepskin.

Chapter 13 is devoted to the elaboration of the technology of biodegradation of fat-containing wastes of meat-processing industry. The researches carried out have shown that the most perspective biodestructor of such substrates is microorganism *Yarrowia lipolytica*. It is established that the preliminary yeast adaptation to metabolites of natural micro flora, the optimum age and concentration of sowing material, carrying out the process at sufficient aeration, pH 5.0-5.5, and the temperature about 30 ^{0}C allow to reduce the duration of cultivation less than 48 hours and to increase the crude protein contents up to 55-60 %.

As described in chapter 14, gas-vortex bioreactor has been created. It uses the absolutely new way of mixing (patents of the USA, Japan, Europe). Practically all types of cells and microorganisms are successfully cultivated in it. Full repeatability of laboratory results at industrial application, low power consumption, mixing of viscous fluids, functioning being filled at 15-90% of volume are its important features. Large-scale production of vaccines in a 300L gas-vortex bioreactor using embryonic cells (CEF) is being launched at the SandP Holding "VIRION".

In chapter 15, nicotinate-phosphoribosiltransferase (NPRT) enzyme was isolated from Brevibacterium ATCC 6872 strain cells that carry out salvage way NAD overproduction. After 500-fold NPRT purification and its basic kinetic characteristics having been found (K_m for nicotinate, ATP and PRPP) the substance was defined as a monomeric protein with the molecular mass of 33.8 – 36.3 kD. The enzyme was found to be completely dependent on ATP ATP and slightly dependent on other nucleosidephosphates. It was discovered to be inhibited by the reaction products: nicotinic acid mononucleotide, pyrophosphate and ADP. Besides, NPPT, obtained from B.ammoniagenes, is subject to retroinhibition by the final product and by the NAD synthesis intermediate metabolite which is its desamidoNAD derivative. NADP does not affect the activity of NPRT produced from B.ammoniagenes ATCC 6872.

In: Industrial Application of Biotechnology
Editors: I. A. Krylov and G. E. Zaikov, pp. 1-8

ISBN 1-60021-039-2

Chapter 1

BIOSENSORIC EXPRESS ANALYZERS FOR SYSTEMS OF AUTOMATIC CONTROL AND MONITORING OF TECHNOLOGICAL PROCESSES IN ALCOHOL PRODUCTION

A. E. Kitova[1,*], A. V. Kuzmichev[2] and A. N. Reshetilov[1]

[1]G. K. Skryabin Institute of Biochemistry and Physiology of Microorganisms RAS, 142290, Pushchino, Russia, Prospect Nauki, 5, Laboratory of Biosensors

[2] "Ecology" Science and Production Center, 117587, Moscow

ABSTRACT

Alcohol oxidase, glucose oxidase, and glucoamylase/glucose oxidase biosensors have been designed to analyze ethanol, glucose and starch concentrations, respectively. The study of parameters of designed biosensors (limits of detection, time and accuracy of analysis, operation and storage stability) leads to a conclusion that they are practically applicable as analyzers for the systems of automatic control and monitoring at different steps of technological process in alcohol production.

Key words: starch, glucose, ethanol, express analysis, enzyme biosensor, microbial biosensor, automatic control systems, automatic monitoring systems, alcohol production technology.

The equipment of basic areas of alcohol production is known to contain mainly sensors for the routine parameters of technological process such as temperature, pressure, levels, and pH, while the concentrations of dissolved starch, sugars and ethanol are currently determined by technical/chemical control methods in winery laboratories. This approach restricts the possibilities of construction and optimization of systems for automatic control and monitoring of technological processes at different areas of alcohol production.

---

*G. K. Skryabin Institute of Biochemistry and Physiology of Microorganisms RAS, 142290, Pushchino, Russia, Prospect Nauki, 5, Laboratory of Biosensors; E-mail: anatol@ibpm.pushchino.ru; fax: +7 (095)-956-33-70

It is also known that the quality of technological processes in alcohol production from starch-containing stuff is characterized actually at all stages by the history dynamics of parameters such as concentrations of dissolved starch, sugars and ethanol. Wide prospects for measuring these parameters are opened by biosensors – analyzers whose action is based on the properties of biological material.

Composition, Action, and Fields of Application of Biosensors

Functionally, a biosensor consists of two main parts: a receptor element (immobilized biological material) and a physicochemical transducer for converting biochemical signal into electric one. Biosensor signal depends on analyte concentration in a sample. Biosensors may be based on enzymes, microbial cells, tissue cultures, DNA, immune components, and organelles. Biosensor signals are recorded by electrochemical (ampero-, potentio- and conductometric), optical, calorimetric and acoustic transducers which register the parameters of biochemical reactions such as appearance of electrochemically active products, temperature changes, intensification or weakening of luminescence [1].

Biosensors are used for continuous control of biochemical processes in biotechnology, determination of foodstuff quality and composition, the content of toxins and antibiotics, as well as ecological monitoring. Achievements in the field of biosensors for food industry have been described in reviews [2, 3]. Enzyme biosensors are commonly used for the monitoring of low-molecular compounds such as glucose, amino acids, and antibiotics. Immunosensors are widely used for detection of pathogenic bacteria, toxins and pesticides in foodstuffs.

Microbial cells are widely used as a basis of bioreceptor. Microbial cells are an available biological material, inexpensive, easily cultivated, and maintained in pure culture. At present, microbial sensors have been designed for detection of sugars, organic acids, alcohols, vitamins, antibiotics, peptides, and inorganic compounds (ammonia, nitrates, nitrites, sulfides, sulfates, phosphates) [4, 5].

Biosensors for Food Industry. Starch, Glucose and Ethanol Detection

Examples of biosensor systems of detection of starch, sugars and ethanol are described below.

Enzyme sensors for ethanol assay may be based on either alcohol dehydrogenase or alcohol oxidase immobilized on an appropriate transducer. Amperometric biosensor for ethanol detection in vapor, based on alcohol dehydrogenase and nicotinamide adenine nucleotide (NAD^+) as a co-factor is presented in [6]. Ethanol detection in vapor is possible within 20-800 ppm. Ethanol biosensor on the basis of alcohol oxidase and oxygen Clark electrode is described in [7]. The range of detection is 0.05 to 10 mM.

Starch detection may be realized by enzyme and microbial sensors. Generally, the scheme of analysis in this case includes starch hydrolysis by amylolytic enzymes (α-amylase, glucoamylase) to glucose followed by glucose detection by amperometric sensor based on glucose oxidase or microbial cells. Moreover, to estimate the total content of utilized sugars in fermentable wort it might be more preferable to use a microbial biosensor, because the broad substrate specificity of microorganisms permits the integral assessment of total sugars [8]. Previously [9], a hybrid membrane-type sensor has been developed on the basis of *Bacillus subtilis* cells and glucoamylase. A glucoamylase membrane was designed to cleave high-molecular substrates (dextrins) to low-molecular products (glucose) which were easily assimilated by microorganisms immobilized on an oxygen electrode.

Basic Requirements for Analyzers of Substance Composition and Concentration in Alcohol Production

The above brief review of biosensors' application in food industry shows that some of them may be successfully used also for solution of scientific and practical problems in alcohol production.

It is also a matter of course that the information about dissolved starch, sugars and ethanol concentrations should come from measuring elements and be processed by a computer in the real time mode. Otherwise, as the process of measuring is long enough, the information lags so much behind that it only shows the already existing state of things but cannot be used for active influence on the technological process [10].

As an example, let us consider the problem of controlling alcohol concentration in brew before distillation. Quite often, the several hours needed for appropriate laboratory analysis result in uncorrectable losses [11]. Similar situations accompany other steps of alcohol production as well.

Instruments for food industry (including alcohol production) must meet the following requirements [10]:

- instruments must be designed by modular approach;
- the action of an instrument must be based on its limit price (benchmark), the cost of a single analysis, and substantiated extreme error of measurements;
- orientation to instruments fixing the kinetics (dynamics) of technological process;
- methods of measurement must be maximally unified

Biosensor analyzers satisfy these principles to a great extent. Besides, they possess the qualities of value for measuring instruments such as high sensitivity, selectivity, and simplicity of implementation.

Biosensor Analyzer "BIOLAN-1010". Performance Specification

One of biosensor analyzers designed with the help of the authors is a control and program (hardware-software) complex including analyzer "BIOLAN-1010" (Fig. 1, A) and PC.

As an example, Table 1 gives specifications of this device for the cases when replaceable modules are bioreceptors – membranes with immobilized biological materials (enzymes or microorganisms) for measurement of dissolved starch, glucose and ethanol concentrations.

The measurement routine implemented in this control and program complex has been described in detail by the authors in [12 - 14].

Table 1.

Characteristics of sensors	Bioreceptor type				
	Alcohol oxidase	*Pichia angusta*	Glucose oxidase	Glucose oxidase and glucoamylase	*Gluconobacte oxydans* and glucoamylase
	ethanol		glucose	starch	
Detection range	0.05 – 5 mM	0.05 – 5 mM	0.05 – 2.50 mM	0.03 – 0.5 g/l	0.03 – 0.5 g/l
Response time, s	60	60	60	200	200
Period of measuring*, min	5	5-10	5	10 - 15	10 - 15
Coefficient of variation, %	2 - 5	3 - 5	2 - 5	3 - 5	3 - 5
Operational stability**, days	7	7	10	10	4
Storage stability (4°C), days	200	30	200	180	90

* Period of measuring includes the time of analysis and the time of signal regeneration

** Sensor stability at continuous measurements

In opinion of the authors, the relative error 2-5%, range of measurements 0.05 to 2.50 mM, operation stability 4 to 10 days and storage stability 90-200 days are quite acceptable for quite a number of objectives of alcohol production, and the period of measurements 5-15 min with the time of analysis 60-200 sec permit the measurements in the real time mode of technological processes of alcohol production. This, in turn, allows biosensor analyzers to be used as sensors for the systems of automatic control and monitoring of technological processes in certain areas of alcohol production [15].

The scheme of starch biosensor assay proposed in [14] may be applied also for assessment of the activity of amylolytic enzymes. Thus, for example, a glucose oxidase (GOD) biosensor was used by the authors to assess the activity of glucoamylase (GA). During the assay, a GA-containing sample (5 μl, 1% solution) and starch solution (100 μl, 1% solution) as a substrate were placed into a 2-ml measuring cuvette of the GOD sensor. The GOD sensor registered the accumulation of glucose formed as a result of starch hydrolysis by amylase; an example of sensor signal registration is shown in Figure 2. Starch hydrolysis was accompanied by sensor current decrease reflecting an increase of glucose concentration in the measuring cuvette. Enzyme activity was expressed in milligrams of glucose formed in a period of time Δt (see Fig. 1) equal to 3 min (this criterion of determining the activity of amylase preparations is taken as a basis and used by different companies (see, e.g., Sigma, 2004-2005). In the example considered, the total activity of glucoamylase component of Alcoholase II-400 was 0.54 U of activity, which in terms of mass units is equal to 10 U of activity/mg of dry mass.

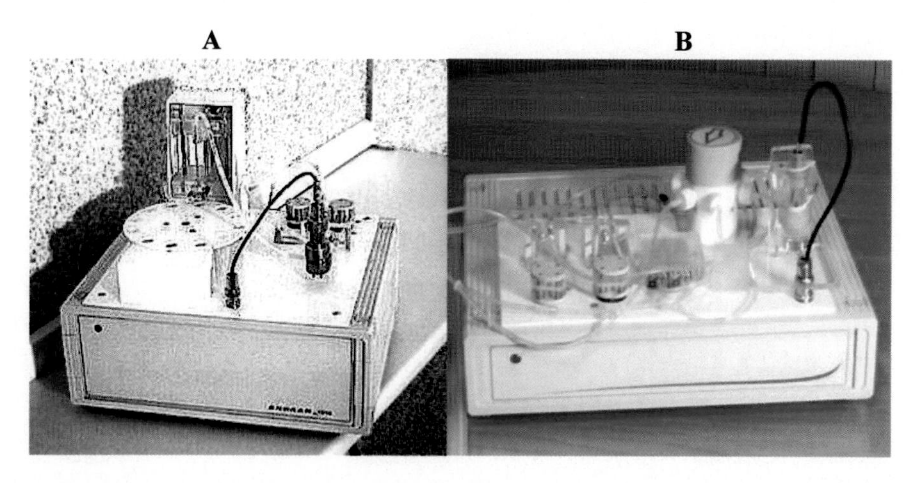

Fig. 1. Physical configuration of biosensor express analyzers: A – "BIOLAN-1010" for starch, glucose, ethanol, and enzyme activity assay; B – "BioRAN-01" for BOD index express assay

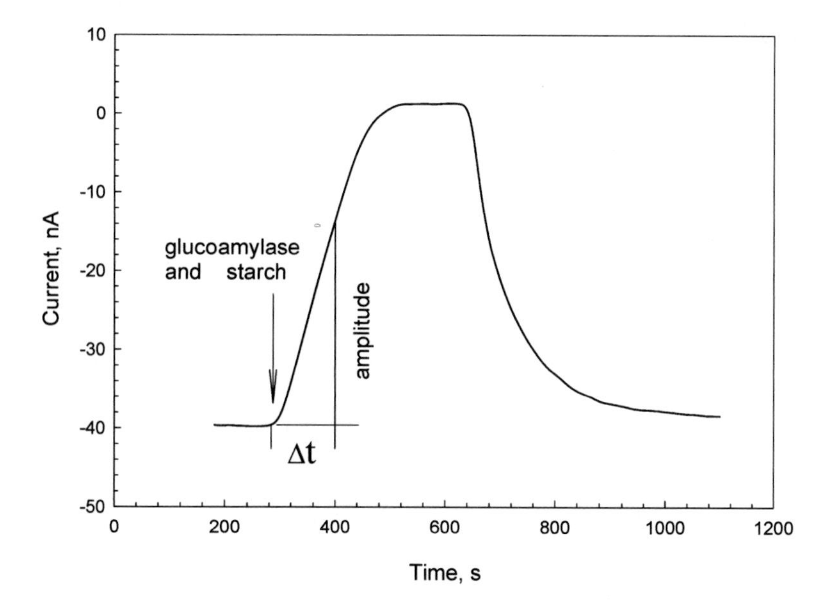

Fig. 2. GOD sensor response to 5 μl of 1% solution of Alcoholase II-400 and 100 μl of 1% starch solution. Cuvette volume is 2 ml

Considering different analytical problems of importance for alcohol production, one might also mention another field of effective application of biosensor assay. Thus, it is known that microbial biosensors are suitable for express analysis of BOD (Biological Oxygen Demand) in wastewaters of different industries, municipal wastewater, etc. [16]. In alcohol production, an urgent problem is assessment of BOD in drainage, because in this case the BOD index is determined both by microflora (yeast cells) and by residual mass of non-fermented sugars and starch present in wastewater. For quick assessment of BOD, it was found efficient to use analyzer "BioRAN-01" developed by the authors – Fig. 1 (B). It is a microbial biosensor of electrochemical type. The time of single sample measurement is no more than 3-5 min, which significantly saves the time and production costs as compared with the conventional BOD_5 method.

It should be particularly mentioned that the changeover to measurement of other substances or other enzyme activities is realized just by replacement of bioreceptor; the method of measurement is the same.

BIOSENSORIC EXPRESS ANALYZER "BIOLAN-1010" FOR ALCOHOL PRODUCTION PROBLEMS

Table 2 shows the main steps of typical technological process of grain-to-alcohol conversion down to rectification, with indication of technological parameters measured by biosensoric express analyzer "BIOLAN-1010", the dynamics of which most completely characterizes the quality of technological process at a given step.

Table 2

##	Stage	Basic technological parameters	Mode of measurement
1	Preparation of grain for processing	Content of starch and sugars in grain	Manual
2	Cooking of starch-containing material	Content of dissolved starch	Automatic
3	Saccharification of starch-containing material	Content of dissolved starch Content of dissolved sugars	Automatic
4	Yeast cultivation	Fermentative quality of yeast biomass	Automatic
5	Fermentation of saccharified mass	Content of dissolved starch Content of dissolved sugars Content of ethanol in brew	Automatic
6	Brew distillation	Residual content of dissolved sugars in brew Content of ethanol in brew Content of ethanol in distilled water-alcohol vapor	Automatic

Additional problems that can be solved using the biosensoric express analyzer "BIOLAN-1010" are as follows:

- determination of activity of enzyme preparations;
- selection of optimal (in the given sense) grades and dosage standards of enzyme preparations at different stages of technological process;
- substantiation of expediency and optimization of parameters of unconventional methods of action on grain and intermediate products, such as vapor-dynamic homogenization, extrusion, different methods of magnetic (electromagnetic) and electric fields;

- selection of optimal temperature regimes at different steps of technological process

As regards "BioRAN-01", it is suitable for express assay of the BOD index in alcohol production wastewater.

Solution of the above problems would result in the rational use of raw material (grain and enzymes) and energy (vapor and electric power) resources and a close approach to development and practical implementation of a system for complex automatization of alcohol production as a whole on the basis of modern computer equipment and technologies.

REFERENCES

[1] Turner et al. (Eds.): *Biosensors: fundamentals and applications*. Oxford University Press, New York, 1987. 770 p.

[2] A. N. Reshetilov: *Prikl. Biokhim. Microbiol.*, 32 (1), 78 (1996). (in Russian).

[3] H. Nakamura, I. Karube : *Anal. Bioanal. Chem.*, 377, 446 (2003).

[4] A. N. Reshetilov, A. V. Lobanov, N. O. Morozova, R. V. Greene, T. D. *Leathers: Sensor Systems*, 12 (4), 486 (1998). (in Russian).

[5] S. F. D'souza: *Biosens. Bioelectron.*, 16, 337 (2001).

[6] J.-K. Park, H.-J. Yee, K. S. Lee , W.-Y. Lee, M.-C. Shin, T.-H. Kim, S.-R. Kim: *Anal. Chim. Acta.*, 390, 83 (1999).

[7] N. O. Morozova, V. V. Ashin, Yu. A. Trotsenko, A. N. Reshetilov: *Prikl. Biokhim. Microbiol.*, 35 (5), 604 (1999). (in Russian).

[8] A. N. Reshetilov, A. E. Kitova, V. A. Alferov, O. N. Ponamoreva, A. V. Kuzmichev, A. A. Ezhkov: *Abstracts of 7-th International Seminar - Presentation of Innovation Scientific and Technical Projects*. November 24 th - 25 th, Pushchino, 2003. P. 176. (in Russian).

[9] R. Renneberg, K. Riedel, P. Liebs, F. Scheller: *Anal. Lett.*, 17 (B5), 349 (1984).

[10] V. M. Balashov, A. A. Begunov: *Food industry*, 2, 68 (2002). (in Russian).

[11] V. I. Sergeyev et al.: *Liquor-vodka production and wine-making*, 5 (29), 3 (2002). (in Russian).

[12] A. E. Kitova, V. A. Alferov, O. N. Ponamoreva, A. V. Kuzmichev, A. A. Ezhkov, T. A. Reshetilova, A. N. Reshetilov // *Enzyme biosensors for glucose, ethanol and starch in fermentation broth* // V. A. Polyakov (Ed.): *Microbial biocatalysts and perspectives of development of enzyme technologies in processing sectors of agroindustrial complex*. Pishchepromizdat, Moscow, 2004. P 255-262. (in Russian)

[13] A. E. Kitova, V. A. Alferov, O. N. Ponamoreva, V. V. Ashin, A. V. Kuzmichev, A. A. Ezhkov, D. V. Arsenyev, A. N. Reshetilov // *Enzyme biosensors for starch, glucose and ethanol concentrations express-analysis* // G. E. Zaikov (Ed.): *Biotechnology and Industry*. Nova Science Publishers, Inc., New York, 2004. P. 19-29.

[14] A. N. Reshetilov, A. E. Kitova, V. A. Alferov, O. N. Ponamoreva, A. V. Kuzmichev, A. A. Ezhkov. Biosensor assay of starch using a glucose oxidase electrode and glucoamylase // Yuryev et al (Eds.): *Starch: from starch containing sources to isolation of starches and their applications*. Nova Science Publishers, Inc., New York, 2004. P. 17-28.

[15] A. E. Kitova, O. N. Ponamoreva, V. A. Alferov, A. V. Kuzmichev, A. A. Ezhkov, D. V. Arsenyev, A. N. Reshetilov: *Liquor-vodka production and wine-making*, 4, 11 (2004). (in Russian).

[16] M. Reiss, A. Heibges, J. Metzger, W. Hartmeier: *Biosens. Bioelect*ron., 13, 1083 (1998).

In: Industrial Application of Biotechnology
Editors: I. A. Krylov and G. E. Zaikov, pp. 9-16
ISBN 1-60021-039-2

Chapter 2

EFFECTS OF LIGHT AND NITROGEN AVAILABILITY ON GROWTH AND ACCUMULATION OF ARACHIDONIC ACID IN THE MICROALGA *PARIETOCHLORIS INCISA*

M. N. Merzlyak[1], A. E. Solovchenko[1], O. B. Chivkunova[1], I. Khozin-Goldberg[2], S. Didi-Cohen[2], and Z. Cohen[2]

[1]Dept. of Physiology of Microorganisms, Fac. of Biology, Moscow State University, 119992, GSP-2 Moscow, Russia

[2]Microalgal Biotechnology Laboratory, The Albert Katz Department of Dryland Biotechnologies, The Jacob Blaustein Institute for Desert Research, Ben-Gurion University of the Negev, Sede-Boker Campus, 84990, Israel

ABSTRACT

Parietochloris incisa is a unicellular, fresh water green alga, capable of accumulating high amounts of the valuable very long chain polyunsaturated fatty acid (VLC-PUFA) arachidonic acid in triacylglycerols (TAG) in cytoplasmic oil-bodies. When growth is retarded by stress conditions, such as high light or nitrogen starvation, the biosynthesis of lipids is enhanced at the expense of protein or carbohydrate production. Thus, under nitrogen starvation, TAG accounted for over 30% of dry weight (DW) and AA content became as high as 60% of total fatty acids. To find out the cultivation conditions providing maximum yield of AA, the effects of light irradiance and N- availability on the DW and AA accumulation have been studied. From the standpoint of biomass accumulation, a light intensity of ca. 400 $\mu E\ m^{-2}\ s^{-1}$ was found to be optimal for growing *P. incisa* on complete medium. Lower light intensities (or higher cell density of inoculum) were found to result in high yield of AA when cultivated on nitrogen-free media. In the absence of nitrogen, algal cells were unable to cope with high light and suffered from photooxidative damage, whereas the nitrogen-sufficient culture survived under such illumination conditions probably due to accumulation of carotenoids. Nitrogen-deprived *P. incisa* cells displayed elevated sensitivity to light.

Keywords: Alga cultivation, arachidonic acid, Parietochloris incisa, biotechnology.

Abbreviations

AA — arachidonic acid;
TAG — triacylglycerols;
TFA — total fatty acids.

Introduction

Very long-chain polyunsaturated fatty acids (VLC-PUFAs) of the ω3 family are quite abundant in microalgae, however, ω6 VLC-PUFA are relatively rare [Cohen *et al.*, 1992]. Arachidonic acid (AA, 20:4ω6) is almost excluded from the lipids of fresh water algae and in most marine species it does not accumulate in the biomass [Bigogno *et al.*, 2002]. When present, VLC-PUFAs are predominantly located in the polar lipids of membranes, whereas triacylglycerols (TAG) generally contain very little VLC-PUFAs [Cohen, 1999]. The ability of cells to accumulate VLC-PUFAs is intrinsically limited in most algae, since these fatty acids are generally components of membrane lipids which content is strictly regulated. Nevertheless, certain algal species can be induced to synthesize and accumulate extremely high proportions of TAG [Shifrin and Chisholm, 1981].

An algal strain has been isolated from the snowy slopes of Mt. Tateyama in Japan and identified as *Parietochloris incisa* comb. nov (Chlorophyta, Trebuxiophyceae) [Watanabe *et al.*, 1996]. This is a unicellular, fresh water green alga, capable of accumulating high amounts of valuable AA-rich TAG in cytoplasmic oil-bodies. The detailed analysis of fatty acid (FA) and lipid composition of *P. incisa* under different environmental conditions have proven it to be one of the richest sources of AA [Bigogno *et al.*, 2002; Khozin-Goldgberg et al., 2002]. AA is the major fatty acid of *P. incisa*, comprising 33.6% of total fatty acids (TFA) in the logarithmic phase and 42.5% in the stationary phase. When the growth of *P. incisa* is retarded under stress conditions, the biosynthesis of lipids is enhanced. Thus, under nitrogen starvation, TAG accounts for over 30% of DW and the proportion of AA becomes as high as 60% of TFA. This finding is of particular practical interest since AA, one of the major FA of brain cell phospholipids [Hansen *et al.*, 1997], has been shown to improve infant development [Koletzko and Brown, 1991].

The optimization of cultivation conditions from the standpoint of both AA and biomass production turned out to be a non-trivial problem [Cheng-Wu *et al.*, 2002]. Thus, low light is limiting for algal growth whereas high light irradiance facilitating rapid accumulation of biomass may cause photodamage, especially under N-limiting condition.

The goal of this study was to determine the influence of light intensity and N-starvation on the production of biomass, TFA and AA in *P. incisa*, in attempt to find out the cultivation conditions providing maximum yield of AA.

EXPERIMENTAL

Cultivation Conditions

A culture of *P. incisa* was isolated from Mt. Tateyama in Japan (Watanabe et al., 1996). The cultures were grown on complete (+N) and N-free (–N) BG-11 medium-(Stanier et al., 1971) in glass columns under constant illumination (by daylight fluorescent lamps) of three different intensities (Table 1) and with constant bubbling of a CO_2–air (1: 99) mixture at 25 °C. Initial chlorophyll (Chl) content was maintained at 30 mg/L in all cases. The inoculum culture was daily diluted to maintain logarithmic growth. For N-starvation, cells were washed three times with sterilized distillated water and resuspended in (–N) BG-11.

Table 1. The experimental design

Conditions	Variant					
	LL*+N	ML+N	HL+N	LL–N	ML–N	HL–N
Nitrogen in the medium	+	+	+	–	–	–
Illumination (μE m^{-2} s^{-1})	35	200	400	35	200	400

*LL — low light; ML —medium light; HL — high light

Fatty Acid and Pigment Analysis

Freeze-dried cells, lipid extracts, or individual lipids were transmethylated with 2% H_2SO_4 in methanol at 80 °C for 1 h. Heptadecanoic acid was added as an internal standard. Gas-chromatographic analysis was performed according to Cohen *et al.* [1993]. Fatty acid methyl esters were identified by co-chromatography with authentic standards (Sigma Co., St Louis) and by comparison of their equivalent chain length [Ackman, 1969]. The data shown represent mean values with a range of less than 5% for major peaks (over 10% of fatty acids) and 10% for minor peaks, of at least two independent samples, each analyzed in duplicate. Chl and Car contents of extracts were measured spectrophotometrically [Wellburn, 1994].

RESULTS AND DISCUSSION

Effect of Light Irradiance and Nitrogen Availability on Biomass Production

As could be seen from Fig. 1, the growth depended on N availability and light irradiance. The culture grown under the strongest light with ample nitrogen (HL+N) showed the highest growth rate and attained the higher biomass in 14 d. The biomass accumulation of ML+N and LL+N cultures was 2–2.5 times lower in comparison to HL +N. By the third day of cultivation on N-free medium, the biomass accumulation of the HL–N culture was higher than that of ML-N and LL-N. However, this culture did not accumulate biomass any further. One should note the similarity in the shape of the growth curves of the ML–N and LL–N cultures. However, ML-N attained highest biomass at the end of cultivation period but it was still lower then the biomass attained by ML+N.

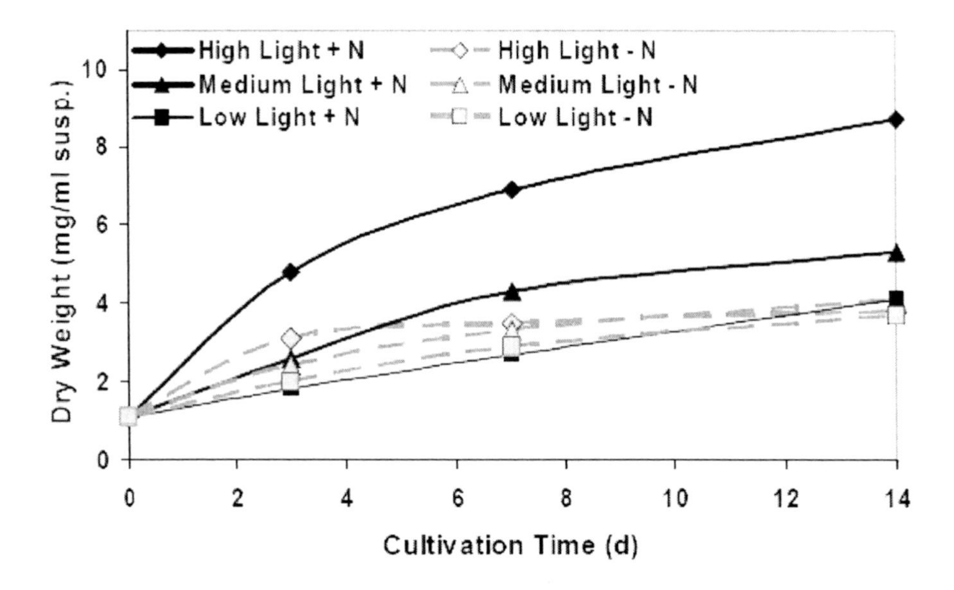

Fig. 1. Effect of light irradiance and nitrogen availability on accumulation of biomass in *P. incisa* cultures

The culture grown on N-free medium under the strongest light demonstrated visual signs of bleaching by the 14th day of the experiment. The N-deprived cultures exposed to high light showed a loss of Chl during the first week, the LL–N culture sustained weak changes in Chl content, whereas the HL–N culture was bleached, probably as a result of photooxidative damage (data not shown).

Both the HL+N and the ML+N, but not the LL+N culture accumulated significant amounts of carotenoids (Car). As a consequence, in the HL+N and ML+N cultures, the Car/Chl ratio increased during the growth period (data not shown). This increase was dependant on the intensity of illumination and was most pronounced in algae cultivated under strong light (HL). N-starved algae accumulated ca. half of the amount of Car but possessed ca. twice higher Car/Chl ratio as compared with N-sufficient cultures grown under the same light intensities.

Effect of light irradiance and nitrogen availability on the fatty acid content and composition in P. incisa cultures.

As can be seen from Table 2, HL triggered lipid accumulation in *P. incisa*, cultivating on complete medium. While nitrogen was initially sufficient, the HL+N culture demonstrated after 14 d the highest accumulation of FA, reaching 35.7% of DW. However, the proportion of AA reached only 38.4% (of TFA) and the AA content was only 13.7% of DW. The ML+N culture had a lower FA content (17.2%), with a higher proportion of AA (46.1%), resulting in an AA content of 7.9% (45% of TFA and 17.2% of DW). One may suggest that the lower proportion of AA in the HL+N stems from photo-oxidative events occurring in the HL+N culture. In the LL+N, *P. incisa* accumulated very little FA or AA (in comparison with the initial conditions). Taking together, these data indicate that, under N-replete conditions, light intensity is an important factor, triggering and regulating AA synthesis in this alga.

Table 2. Effect of nitrogen availability and light irradiance on the fatty acid composition and content (% of DW) in *P. incise*

Variant	Days*	16:0	18:0	18:1 ω9	18:2	18:3 ω6	18:3 ω3	20:3 ω6	20:4 ω6	AA % DW	TFA % DW
HL+N	3	14.0	3.0	18.6	19.1	2.1	3.6	1.7	27.6	4.2	15.4
	7	9.7	2.6	22.3	15.6	1.1	1.7	1.5	36.5	9.0	24.6
	14	8.6	2.2	23.9	15.8	0.7	1.2	1.0	38.4	13.7	35.7
HL–N	3	11.4	2.0	25.0	12.5	1.6	1.6	1.1	36.5	6.8	18.6
	7	10.4	2.5	19.1	11.8	0.8	1.0	1.0	46.4	12.2	26.7
	14	9.5	2.2	14.5	10.9	0.7	0.7	0.9	52.2	15.1	28.9
ML+N	3	16.8	1.9	7.5	25.3	0.9	4.8	0.4	28.1	3.7	13.3
	7	15.2	2.3	7.7	23.7	1.1	2.7	0.7	32.2	3.7	11.7
	14	11.5	2.8	10.5	16.6	0.8	1.0	1.1	46.1	7.9	17.2
ML–N	3	13.2	2.4	11.0	13.3	1.6	2.1	1.4	44.4	6.7	15.1
	7	11.6	3.0	11.9	11.5	1.0	1.1	1.3	54.6	13.2	25.1
	14	10.4	2.7	11.1	9.7	0.7	0.7	1.2	56.5	19.0	33.5
LL–N	3	17.9	1.5	7.1	22.3	0.8	5.1	0.6	26.3	2.3	8.6
	7	16.8	1.7	7.7	23.3	0.5	3.5	0.5	24.9	2.2	8.7
	14	14.9	2.0	7.5	23.9	0.8	2.1	0.7	30.9	3.4	11.0
LL–N	3	13.5	2.4	7.3	17.7	1.7	2.9	0.9	41.4	5.3	12.8
	7	10.7	2.8	8.0	12.4	1.0	1.1	1.2	55.0	11.8	21.5
	14	10.4	2.5	7.9	10.7	0.7	0.8	1.2	58.2	14.9	25.6

*TFA content of inoculum (day 0) — (8 % of DW, including 1.8% AA)

This is in accord with the results of outdoor experiments by Cheng Wu *et al.* [2002] showing that irradiation greatly affected the culture volume content of AA, seemingly due to increased cell biomass under high light. The cumulative culture content of AA after 38 culture days in high light was ca. twice as higher compared with the low light treatment. The highest AA-proportions of TFA obtained in the laboratory were 58 and 56 %, under low and high light, respectively.

In the absence of nitrogen, all cultures enhanced their TFA accumulation. The highest TFA (33.5%) and AA content (19%) were observed in the ML-N culture. The FA content of the HL-N culture increased only for the first 7 d and did not change appreciably thereafter. The LL–N culture demonstrated the highest proportion of AA (58%), but the TFA (25.6%) and AA (14.9%) content were substantially lower than in the ML-N culture.

A linear interdependence was found between the Car/Chl ratio and the volumetric content of AA in the ML-N and HL-N cultures (Fig. 2). The strong correlation between the two parameters also existed for the HL-N culture, but in a different range of Car/Chl ratios, because photo-damaged and bleached HL-N culture contained significantly more carotenoids. Thus, this parameter can be utilized for an estimation of the volumetric content of AA in the N-starved cultures, cultivated in the range of low (35) to relatively high (200 $\mu E\ m^{-2}\ s^{-1}$) light intensities under laboratory conditions.

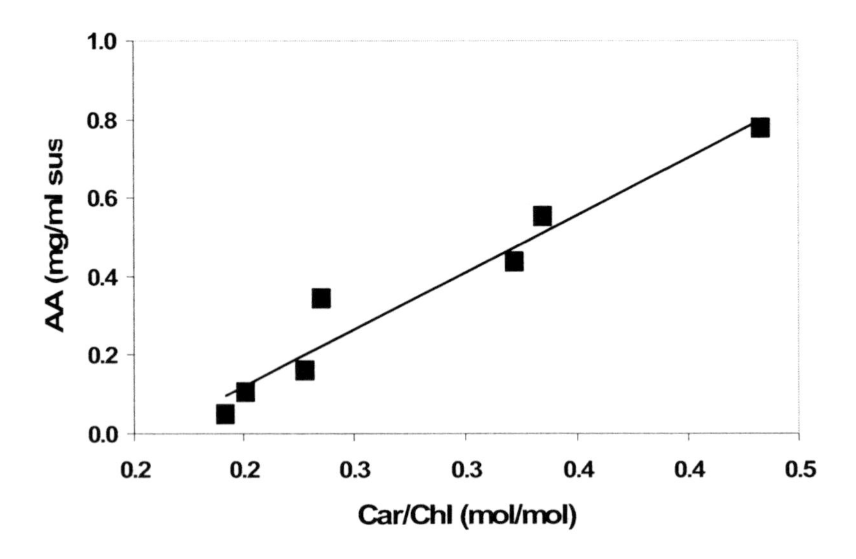

Fig. 2. Correlation between the Car/Chl ratio and the volumetric content of AA in *P. incisa* cultures (NL-N and LL-N)

In many microalgal species, stresses such as nutrient deficiency and high light lead to significant induction of lipid biosynthesis to provide additional sink for photosynthates [Thompson, 1996]. Possible physiological meanings of this phenomenon include also the facilitation of the accumulation of carotenoids, a mechanism found in *Dunaliella salina* [Mendoza *et al.*, 1999], *Haematococcus pluvialis* [Boussiba 2000; Zhekisheva *et al.*, 2002], which may serve for photoprotection [Wang *et al.*, 2003; Zhekisheva *et al.*, 2002]. Our preliminary data indicate that accumulation of carotenoids (namely, lutein and β-carotene), especially pronounced under high light and N-deficient conditions, also takes place in whole cells and oil bodies of *P. incisa*.

CONCLUSION

Collectively, the results obtained in this work suggest that:

- Light intensity is an important factor affecting AA production in *P. incisa* under nitrogen starvation.
- From the standpoint of biomass accumulation, light intensities of about 400 μE m^{-2} s^{-1} are optimal for growing *P. incisa* on complete medium. Lower light intensities (or higher cell density of inoculum) should be considered for cultivation on media lacking nitrogen.
- In the absence of nitrogen, algal cells were unable to cope with high light (~400 μE m^{-2} s^{-1}) and suffered from photooxidative damage; the nitrogen-sufficient culture survived under such illumination conditions.

- Nitrogen-deprived *P. incisa* cultures display elevated sensitivity to light intensity, the lack of nitrogen limits the (photo) adaptive potential of the alga, probably, by reducing protein synthesis capacity.

ACKNOWLEDGEMENTS

This work was supported in part by fellowships from the BIDR to MNM and AES.

REFERENCES

[1] Ackman R.G. (1969) Gas–liquid chromatography of fatty acids and esters. In: Lowenstein, J.M. (Ed.), *Method. Enzymol.* Vol. 14. Academic Press, New York, pp. 329–381.

[2] Bigogno C. (2000). Biosynthesis of Arachidonic Acid (AA) in the Microalga *Parietochloris incisa* and the Effect of Environmental Conditions on the Production of AA. PhD thesis, Ben Gurion University, Israel.

[3] Bigogno C., Khozin-Goldberg I., Boussiba S., Vonshak A., Cohen Z. (2002) Lipid and fatty acid composition of the green oleaginous alga *Parietochloris incisa*, the richest plant source of arachidonic acid. *Phytochemistry* 60: 497–503

[4] Boussiba S. (2000). Carotenogenesis in the green alga Haematococcus pluvialis: cellular physiology and stress response. *Physiol. Plant.* 108: 111–7.

[5] Cheng-Wu Z., Cohen Z., Khozin-Goldberg I., Richmond A. (2002) Characterization of growth and arachidonic acid production of *Parietochloris incisa* comb. nov (Trebouxiophyceae, Chlorophyta). *J. Appl. Phycol.* 14: 453–460, 453.

[6] Cohen Z. (1999). Production of polyunsaturated fatty acids by the microalga *Porphyridium cruentum*. In: Cohen, Z. (Ed.), *Production of Chemicals by Microalgae.* Taylor and Francis, London, pp. 1–24.

[7] Cohen Z., Didi S., Heimer Y. M. (1992) Over-production of γ-linolenic and eicosapentaenoic acids by algae. *Plant Physiol.* 98: 569–72.

[8] Cohen Z., Vonshak A., Richmond A. (1988) Effect of environmental conditions on fatty acid composition of the red alga *Porphyridium cruentum*: correlation to growth rate. *J. Phycol.* 24: 328–332.

[9] Cohen, Z. (1990) The production potential of eicosapentaenoic and arachidonic acids by the red alga *Porphyridium cruentum*. *J. Am. Oil Chem. Soc.* 67: 916–920.

[10] Hansen J., Schade D., Harris C., Merkel K., Adamkin D., Hall R., Lim M., Moya F., Stevens D., Twist P. (1997). Docosahexaenoic acid plus arachidonic acid enhance preterm infant growth. *Prostaglandins, Leukotriens, Essential Fatty Acids* 57: 157.

[11] Koletzko B., Braun M. (1991) Arachidonic acid and early human growth: is there a relation? *Ann. Nutr. Metabol.* 35: 128–131.

[12] Mendoza H., Martel A., Jimenez del Rio M., Garcia Reina G. (1999) Oleic acid is the main fatty acid related with carotenogenesis in *Dunaliella salina*. *J. Appl. Phycol.* 11: 15–9.

[13] Shifrin N.S., Chishlom S.W. (1981). Phytoplankton lipids: interspecific differences and effects of nitrate, silicate, and light–dark cycles. *J. Phycol.* 17: 374–384.

[14] Thompson Jr. G.A. (1996). Lipids and membrane function in green algae. *Biochim. Biophys. Acta* 1302: 17–45.

[15] Wang B., Zarka A., Trebst A., Boussiba S. (2003) Astaxanthin accumulation in *Haematococcus pluvialis* (Chlorophyceae) as an active photoprotective process under high irradiance *J. Phycol.* 39: 1116–24.

[16] Watanabe S., Hirabashi S., Boussiba S., Cohen Z., Vonshak A., Richmond A. (1996) *Parietochloris incisa* comb. Nov. (Trebuxiophyceae, Chlorophyta). *Phycol. Res.*: 44, 107–8.

[17] Wellburn A.R. The spectral determination of chlorophylls a and b, as well as total carotenoids, using various solvents with spectrophotometers of different resolution. *J. Plant. Physiol.* (1994) 144: 307-13.

[18] Zhekisheva M., Boussiba S., Khozin-Goldberg I., Zarka A., Cohen Z. (2002) Accumulation of oleic acid in *Haematococcus pluvialis* (Chlorophyceae) under nitrogen starvation or high light is correlated with that of astaxanthin esters. *J. Phycol.* 38: 325–31.

In: Industrial Application of Biotechnology
Editors: I. A. Krylov and G. E. Zaikov, pp. 17-28
ISBN 1-60021-039-2

Chapter 3

THE IDENTIFICATION OF DEHYDROSHIKIMIC ACID, ACCUMULATED IN FERMENTATION BROTHS. THE USE OF CAPILLARY ELECTROPHORESIS FOR DETERMINATION OF DEHYDROSHIKIMIC ACID IN ENZYME REACTION MIXTURES FOR MEASURING OF ACTIVITY OF SHIKIMATE DEHYDROGENASE

A. E. Novikova[a,*]***, T. A. Yampolskaya***[a]***, K. F. Turchin***[b]***,***
O. S. Anisimova[b] ***and M. M. Gusyatiner***[a]

[a] Closed Joint Stock Company "Ajinomoto-Genetika Research Institute" (CSC "AGRI"), Moscow, Russia
[b] Federal State Unitary Concern "Centre of Chemistry of Drugs-All-Russian Research Chemical-Pharmaceutical Institute", Zubovskaya st.7, 119815 Moscow, Russia

ABSTRACT

The identification of dehydroshikimic acid accumulated in fermentation broths was carried out with the use of different analytical methods. The capillary electrophoresis was used for determination of dehydroshikimic acid in fermentation broths and reaction mixtures of enzyme reaction. These experiments showed, that *B. subtilis* I-118/pDP4 strain had the best shikimate/dehydroshikimate ratio.

Keywords: high-performance liquid chromatography, capillary electrophoresis, indirect detection, electroosmotic flow, shikimate dehydrogenase, shikimic acid, dehydroshikimic acid, Bac. subtilis.

---

[*] Closed Joint Stock Company "Ajinomoto-Genetika Research Institute" (CSC "AGRI"), 1st Dorozhny pr.1, 117545 Moscow, Russia Corresponding author. E-mail address: anovikova_agri@yahoo.com

1. INTRODUCTION

Shikimic acid is the material used in pharmaceutics. This substance is an intermediate in the aromatic amino acids pathway of microorganisms and this possibility can be used to obtain a bacterial producer or a microorganizm producing shikimate. The measurement of the quantity of the main product is the first task in the process of breeding of the producers. The determination of different by-products and precursors of shikimic acid, accumulating in fermentation broth and indicating of certain metabolic disorders is the next task. That is necessary to improve the afficiency of the selection process. Spectrophotometry and high-performance liquid chromatography (HPLC) are widely used for determination of shikimic acid. Most of routine methods are based on the treatment of shikimic acid with periodate in strong acid solution, followed by the reaction of the oxidated product with thiobarbituric acid [1,2]. Optical densities at 535nm are determined in the cyclohexanone phase, after the procedure of extraction. This measurement has a very high sensitivity - 0.01 micromole of added shikimic acid. The method is relatively nonspecific and unreliable to measure this compound in fermentation broths because of high background optical density. In comparison with spectrophotometric method, HPLC with gradient elution or isocratic elution and diode-array detection [2-6] has advantages for analysis both shikimic acid and metabolic precursors at one run. For this investigation the HPLC method was used as the major one.

On the key stages of selection process it is neccesary to determine the activities of enzymes of biosynthetic pathway. As it can be seen from the figure 1, the main enzyme involved in shikimate biosynthetic is shikimate dehydrogenase (EC 1.1.1.25). The shikimate dehydrogenase enzyme activity is ussualy estimated spectrophotometrically at 340nm, in the direction of shikimate oxidation by following the production of NADPH [10]. But this method is used for determination of activity of purified enzymes and can face some problems of interference by other components from cell-free extracts. In this case, the problem can be solved by using the alternative method to confirm the correctness of the data obtained by the classic spectrophotometric methods. Besides, the commercial substrate standard for shikimate dehydrogenase is absent. Therefore, dehydroshikimic acid must be prepared chemically [11] using enzyme synthesis [11-12] or purification from dried cells [12]. It was also expedient to use other methods in addition to spectrophotometric one to check the purity of obtained substrates.

Some authors reported the use of capillary electrophoresis (CE) with indirect detection as an efficient technique for separation of anions of organic acid [13-15], including shikimate [16]. Characteristics like weak influence of protein matrix on capillary lifetime and on the efficiency of separation make CE attractive for using this method for the analysis of the reaction mixtures after enzyme incubation. The method with indirect UV detection [17] was used for the determination of investigated compounds. The absorbing electrolyte contained benzoic acid and Tris. The simplest way to shorten the time of analysis is to change the surface characteristics of the capillary. It was achieved by using an osmotic flow modifier in the buffer, namely tetradecyltrimetylamonium bromide [18]. Thus, CE was used to solve some tasks, as: the identification and the control of purity of prepared substrate and the verification of data of enzyme activities.

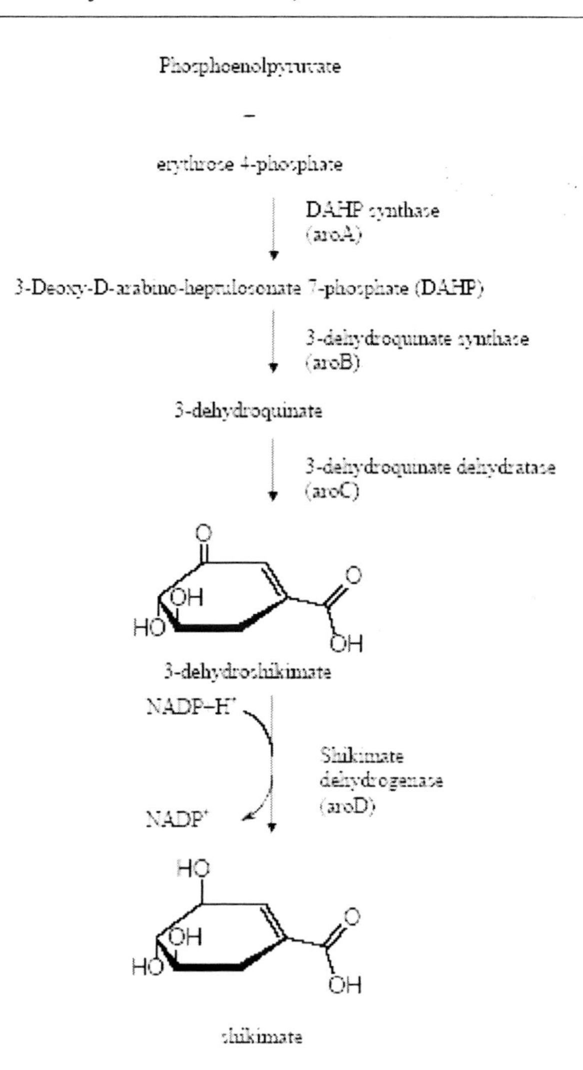

Fig. 1. The scheme of biosynthesis of shikimic acid in Bac.subtilis

2. Experimental

2.1. Materials and Reagents

Potassium chloride, phosphoric acid (85%), hydrochloric acid (37%), glacial acetic acid, triethylamine, Tris (base), NADP sodium salt, NADPH sodium salt, shikimic acid, tetradecyltrimethylammonium bromide, benzoic acid were obtained from Sigma (st. Louis, MO, USA). All solutions were prepared using Milli-Q Academic water (Millipore, Milford, WA, USA).All standard solutions were stored at -20^{0}C before use.

2.2. Bacterial Strains and Plasmids

For our studies we used Bacillus subtilis strains B-1403 (aroD120, lys1, trpC2, recE4). Strain was obtained from Russian National Collection of Industrial Microorganisms (VKPM). Strain Bacillus subtilis I-118** (aroI116, amy4, recE4) was developed by *Dr. S.Yuku* [19].

Plasmids used: pBAG7** [20] has genes aroAG from chromosome of Bacillus subtilis 168, pDP4** [20] has gene aroD from chromosome of Bacillus amyloliquefaciens. Plasmid strains of B.subtilis were obtained by plasmid DNA transformation [21].

2.3. Preparation of Cell-Free Extracts

Bacillus subtilis cells of all strains were cultivated in M9 minimal salt-glucose Spicaizen medium (with addition of necessary amino acids to 50-100 mg/l final concentration for auxotrophic strains) up to the end of log-phase. The cells were harvested by the centrifugation, washed by 0,9 % NaCl solution and resuspended in correspondent buffer for enzyme reaction. The cell suspension was disrupted by sonication (MSE, UK) with amplitude 8-10 mk for 3 minutes at $+4^0$C. Cell debris was removed by centrifugation at 13000 rpm for 15 minutes, $+4^0$C. Obtained supernatant (further cell-free extract) was used for measurement of enzyme activities. The measurement of protein concentration was carried out using the standard Bredford technique (Bio-Rad protein assay KitII).

2.4. Preparation of 3-Dehydroshikimate

The purification of this substance was based on the known methods [11,12]. In this study the other ion-exchange columns were used. 3-Dehydroshikimate is readily accumulated by cells of E.coli or Bac. subtilis with corresponding metabolic blocks in the biosynthetic pathways leading to shikimic acid. After centrifugation, the supernatant solution was passed through a 1×15 cm column of Dowex 50W-X2 (H^+ form) 200-400 mesh. The column was washed with distilled water, 3-ml fractions were collected. The quantity of primary purified substances was estimated by 2,4-dinitrophenylhydrazine reaction [22]. Combined fractions containing the highest amount of target substance were further purified by ion-exchange chromatography on a 1×15 cm column of Dowex AG1-X10 (Cl^- form) 200-400 mesh. 3-Dehydroshikimate is eluted from the column with linear gradient (250ml+250ml) of water and 1M potassium chloride. 3-ml fractions were collected. The quantity of purified substance was also estimated at 234nm [23,24]. All procedures were carried out at $+4^0$C.

2.5. The Assay of Enzyme Activities

Reactions were initiated with adding of 5-10 µl of cell-free extracts up to 50 µl of volume of reaction mixtures and were incubated during 10-15 min at 37^0C. The reactions were stopped by cooling at 0^0C.

The reaction mixture for direct shikimate dehydrogenase enzyme reaction contained: 100mM Tris-HCl buffer (pH 7,5), 1,26mM NADPH, 1mM 3-dehydroshikimic acid. The reaction mixture for shikimate dehydrogenase reaction in reverse direction contained: 100mM Tris-propanol (pH 8,5), 0,26mM $NADP^+$, 1mM shikimic acid [10].

2.6. Spectrophotometric Measurements

The UV1201 model (Shimadzu, Japan) was used for spectrophotometric measurement. These measurements were carried out according to methodologies, described in studies [7-10].

2.7. HPLC

HPLC investigations were carried out using an analytical system consisting of a 510 HPLC pump (Waters, Milford, MA, USA), manual injector valve with a 10µl loop (Rheodyne,USA) and 991 photodiode array detector (Waters, Milford, MA, USA) or 481 Lambda-Max variable wavelength UV-VIS detector (Waters, Milford, MA, USA). Raw data were collected and processed by the 990+ PDA software (Waters, Milford, MA, USA) or ChromandSpec chromatography software (Ampersand Ltd.,RF) accordingly.

For chromatographic separation in reverse-phase mode, Separon (Tessek, Czechia) SGX C_{18} column (150mm in length, 3.3mm inner diameter, 5µm particle size) was used. Isocratic chromatographic conditions with a mobile phase of 0,5%(v/v) glacial acetic acid adjusted to a pH value of 6,0 with a triethylamine were used. The flow was 0.32 ml/min at room temperature. Spectra for the UV library were recorded at a wavelength range from 190 to 300 nm with a spectral resolution of 1.2 nm.

For chromatographic separation in ion-exclusion mode, Ultron (Shinwa Chemical Industries, LTD, Japan) PS-80H W type column (300mm in length, 8mm inner diameter, 10µm particle size) was used. H_3PO_4 (pH 2,0) was used as a mobile phase. The flow was 1 ml/min at room temperature. The wavelength was set at 210nm. These conditions were used for micropreparative purification of 3-dehydroshikimic acid from fermentation broth. This preparation was used for carrying out the 2,4-dinitrophenylhydrazine reaction and obtaining of mass spectra and NMR spectra. 0,05-5 mM aqueous solutions of shikimic acid and dehydroshikimic acid were injected for HPLC separation.

2.8. Capillary Electrophoresis

All electropherograms were generated using a Quanta 4000E system (Waters, Milford, MA, USA) with detector measuring at 254nm. Data were collected and processed using a ChromandSpec (Ampersand Ltd.,RF) chromatography software. Fused silica capillaries (Waters, Milford, MA, USA) of 75 µm inner diameter and 60 cm total length were used. The separation buffer contained 50 mM Tris (basic), 25 mM benzoic acid, 0.25 mM tetradecyltrimethylammonium bromide, pH 8.5. The applied constant voltage was 25kV using the negative power supply. Samples were entered hydrostatically during 25 s. Temperature of the capillary was +18^0 C. 2-40 μM aqueous solutions of shikimic acid, dehydroshikimic acid, dehydroquinic acid and DAHP were injected for CE separation.

2.9. Mass-Spectrometer

The mass spectra were obtained on the mass-spectrometer SSQ-710 (Finnigan, USA). Mass-spectra of electron impact (EI): the energy of ionizing electrons was 70eV, the ion source temperature was 150°C. Direct inlet of the samples in the ion source was used. The inlet system temperature was changed from 20° to 350°C (the speed 2.7 °C/sec). Mass-spectra of chemical desorption ionization (DCI): isobutane was used as reactant gas, the samples were heated up to 800°C with the speed 65°C/sec.

2.10. *NMR*

The NMR spectra were recorded on UNITY+400 spectrometer (Varian, USA), working frequency 400MHz on ^{1}H nucleus. D_2O was used as a solvent, the signal of residual protons of the solvent (HOD, δ 4.67) was the internal standard.

3. Results and Discussion

3.1. Identification of 3-Dehydroshikimic Acid

During the routine analysis of shikimic acid by reversed-phase HPLC method, one more component were detected in fermentation broth of strain I-118/pBAG7 with defective shikimate kinase (EC 2.7.1.71). As it is shown on the figure 2, this component had high response and maximum absorbance at 234nm. Moreover, the analysis of the fermentation broth of another strain B-1403 with defective shikimate dehydrogenase also showed the presence of this unknown compound (Fig. 2b). For more detailed investigation the samples were analyzed by ion-exclusion chromatography with longer run time, but better resolution than it was in reversed-phase chromatography regime. In this case also two major peacks were detected (Fig. 2c). Thus, obtained chromatograms and UV spectra allowed to assume that this substance can be the precursor of shikimic acid, namely 3-dehydroshikimate (Fig. 1). Then, this compound was isolated from fermentation broth with highest amount of it and qualitative reaction with 2,4-dinitrophenylhydrazine confirmed the presence of keto-group.

This supposition was finaly confirmed by mass-spectrum and NMR-spectrum obtained for isolated compound. For mass-spectra study of the purified substance, the ionization by EI and DCI were used. Firstly, the mass-spectra of shikimic acid standard were investigated. In the EI spectrum, the peaks 156 $[M-H_2O]^+$, 138 $[M-2H_2O]^+$, 115 $[M-H_2O-CHCO]^+$, 110 $[M-H_2O-HCOOH]^+$, 97 $[M-2H_2O-CHCO]^+$, 60 $[HOCH=CHOH]^+$, 53 $[C_4H_5]^+$ and low-intensity (< 1 %) peaks of ions 174 (M^+) and 175 (MH^+) are observed. Also, the intensive peak of MH^+ ion (m/z = 175) is observed in the DCI spectrum of shikimic acid. The EI mass spectrum of the sample containing 3-dehydroshikimate presumably is characterized by peak of a molecular ion (m/z = 172). The DCI mass-spectrum of this sample is characterized by intensive peak of MH+ ion (m/z = 173). In the EI spectrum the peaks of ions 154 $[M-H_2O]^+$, 137 $[M-H_2O-OH]^+$, 127 $[M-COOH]^+$, 126 $[M-HCOOH]^+$, 109 $[M-H_2O-COOH]^+$, 81 $[M-H_2O-COOH-CO]^+$ are determined. According to mass spectrometric fragmentation, the molecule of researched substance containes COOH-group and, as minimum, one OH – group.

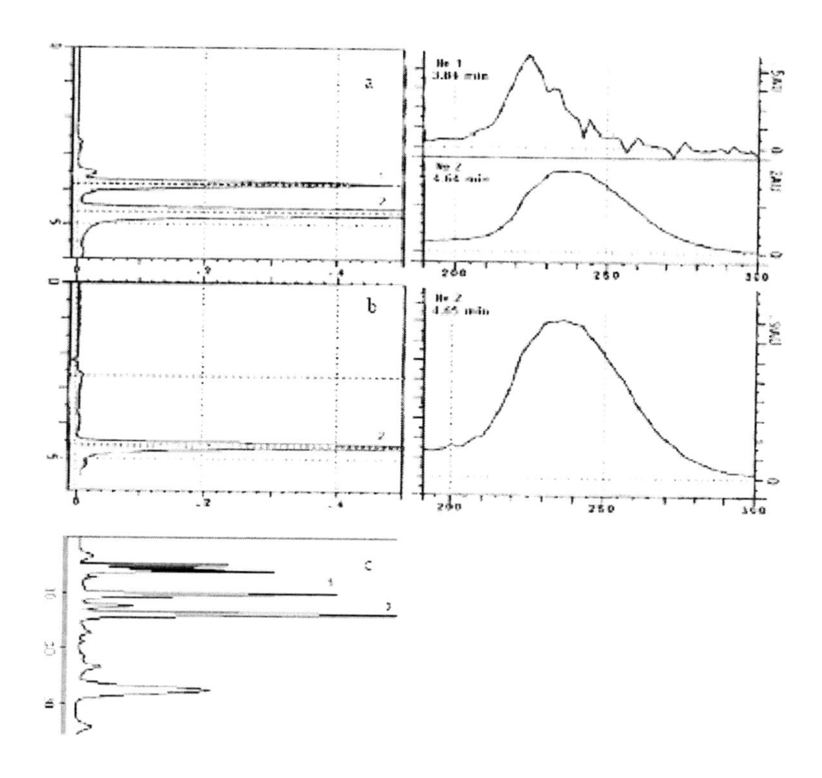

Fig. 2. Chromatograms of fermentation broths: a) Bac. subtilis I-118/pBAG7 strain (reverse-phase HPLC); b) Bac. subtilis aroD B-1403 strain (reverse-phase HPLC); c) Bac. subtilis I-118/pBAG7 strain (ion-exchange HPLC). Dilution was 20 times. 1= shikimate; 2= 3-dehydroshikimate

Position and orientation of the functional groups in the purified substance were elucidated by means of ^1H NMR spectroscopy. The spectral parameters for the substance are presented in the table 1. The high absolute value of the spin coupling constant $^2J_{6A,6B}$ ~ 18 Hz in the purified substance spectrum (as well as in the shikimate spectrum) indicates the juxtaposition of $C_6H_AH_B$ group to C=C double bond. The observed vicinal couplings $^3J_{5,6A}$=5.3 Hz, $^3J_{5,6B}$=10.1 Hz and $^3J_{4,5}$=11.2 Hz demonstrate pseudo-diaxial (trans) orientation of H_5 and H_4 in the purified substance; they prove also that only C_3 could be oxidized to carbonyl in the compound. It agrees well with the absence of any vicinal couplings for H_2 and its long-range coupling to one of 6-H ($^4J_{2,6B}$=3.1 Hz).

Thus, the data received by NMR and mass spectrometric techniques give grounds to consider that the purified substance is 3-dehydroshikimic acid. Hence the definite proof of accumulation of this precursor in fermentation broths of some mutant strains has allowed to assume that the shikimate dehydrogenase as an enzyme catalysing the transforming of 3-dehydroshikimate into shikimate was the bottleneck in obtained strains.

Table I. Chemical Shifts (δ, ppm) and Spin Coupling Constants (J, Hz) in ^{1}H NMR Spectra of 3-Dehydroshikimic and Shikimic Acids

	2-H	3-H	4-H	5-H	6-HA	6-HB
3-dehydro-shikimic acid	6.17d $J_{2,6B}$=3.1		4.03d $J_{4,5}$=11.2	3.76m $J_{5,6A}$=5.3 $J_{5,6B}$=10.1	2.82q $J_{6A,6B}$=- 18.1	2.41m
shikimic acid	6.68m $J_{2,3}$=4.0 $J_{2,6A}$=1.7 $J_{2,6B}$=2.0	4.31m $J_{3,4}$=4.4	3.64q $J_{4,5}$=8.2	3.90m $J_{5,6A}$=5.2 $J_{5,6B}$=6.4	2.60m $J_{6A,6B}$=-18.2	2.09m

According to this, the shikimate dehydrogenase activity had to be enhanced, for example, by increasing copy number of a gene encoding shikimate dehydrogenase (aroD). For this aim, the gene aroD was cloned from *Bacillus amyloliquefaciens* strain, that allowed to enhance the expression of *aroD* gene, thereby enhancing shikimate dehydrogenase activity. Obtained *B. subtilis* strains were cultivated and shikimic acid and dehydroshikimic acid in the fermentation medium were analyzed as described above. The result is shown in the table 2. As it can be seen (table 2), *B. subtilis* aroI116 strain which carried pDP4 plasmid produced highest quantity of shikimate (around 14 g/l) and lowest amount of dehydroshikimate.

Table II. Accumulation of 3-Dehydroshikimic (DHSH) and Shikimic (SH) Acids in fermentation broths of different strains

Strain	Time Hr	Shikimate g/l	Dehydro-Shikimate g/l	Ratio DHSH/SH
I-118	24	3.1	4.8	1.5
	45	5.9	8.0	1.3
	70	8.5	9.5	1.1
I-118 /pDP4	24	6.5	1.9	0.29
	45	14.0	3.6	0.25
	70	14.0	6.8	0.48
aroI118 /pBAG7	24	4.9	14	2.8
	45	8.0	24	3.0
	70	7.2	20	2.7

3.2. The Determination of 3-Dehydroshikimate by Capillary Electrophoresis

To determine the activity of shikimate dehydrogenase both in reverse and forward directions it was necessary to prepare purified 3-dehydroshikimate. For this purpose, the fermentation broth containing the largest amount of 3-dehydroshikimate was chosen, then it was passed through two preparative columns. Collected fractions were analysed by both spectrophotometric and electrophoretic methods. In that case the spectrophotometric measurement, namely determination of 2,4-dinitrophenylhydrazone of 3-dehydroshikimic acid, gave an opportunity to make a primary selection of the fractions, which contained the maximum amount of substrate. The purity of these fractions was estimated by CE. The background electrolyte (25mM benzoic acid, 50mM Tris, 0,25mM TTAB) for indirect detection was used to observe the most of anion components of fractions selected on the previous step with 2,4-dinitrophenylhydrazine reaction. As it can be seen from figure 3, 3-dehydroshikimate was partially purified on the first ion-exchange column and CE analysis allowed to choose the fractions without shikimate.

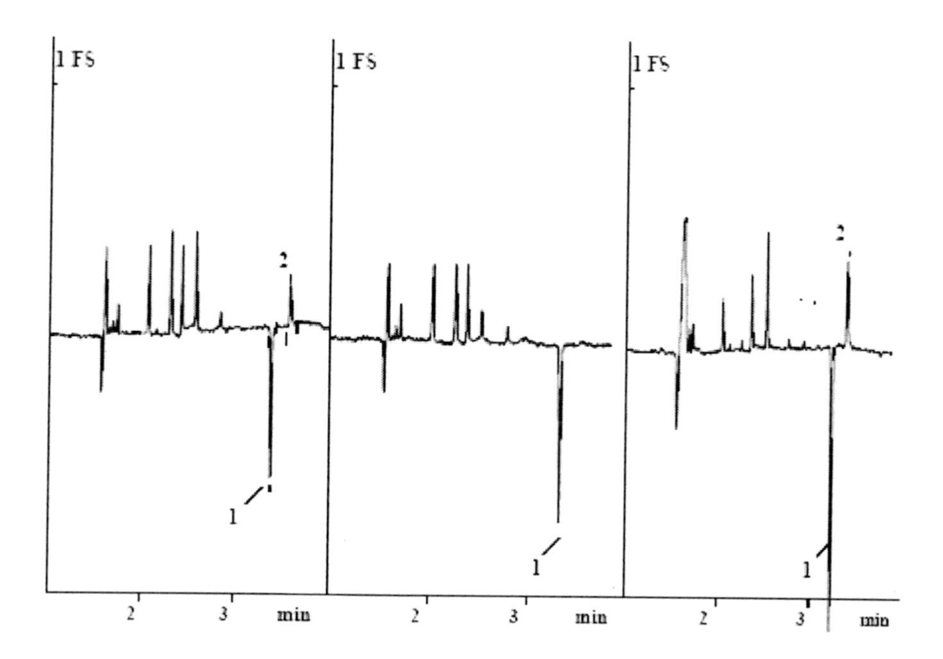

Fig. 3. Electropherogramms of purified fractions after separation on Dowex 50W-X2 resin a) № 3; b) № 4; c) accumulative medium of strain Bac. subtilis I-118/pBAG7. Dilution was 1010 times. 1= 3-dehydroshikimate; 2= shikimate

After that, the crude 3-dehydroshikimate selected on the first stage was purified on the second ion-exchange column. The checking of these fractions by CE gave possibility to choose the sample purified from main impurities (Fig. 4).

This fraction was chosen as a substrate for shikimate dehydrogenase enzyme. Mainly this enzyme was assayed spectrophotometrically by the measuring of changing in optical density at 340nm. However, this study was carried out with crude or partially purified extracts and the some background activity of NADPH oxidases was observed. Therefore, there was a need to confirm the spectrophotometric data using additional alternative method. For this aim CE was applied. As can be taken from figure 5, the electropherogram shows the process of formation of shikimate upon the time. In its turn, the formation of 3-dehydroshikimate was observed (Fig. 6) in condions of reverse enzyme reaction. This experiment showed, that *B. subtilis* I-118 strain which carries pDP4 plasmid had 50 times higher activity of shikimate dehydrogenase than the strain without plasmid (table 3).

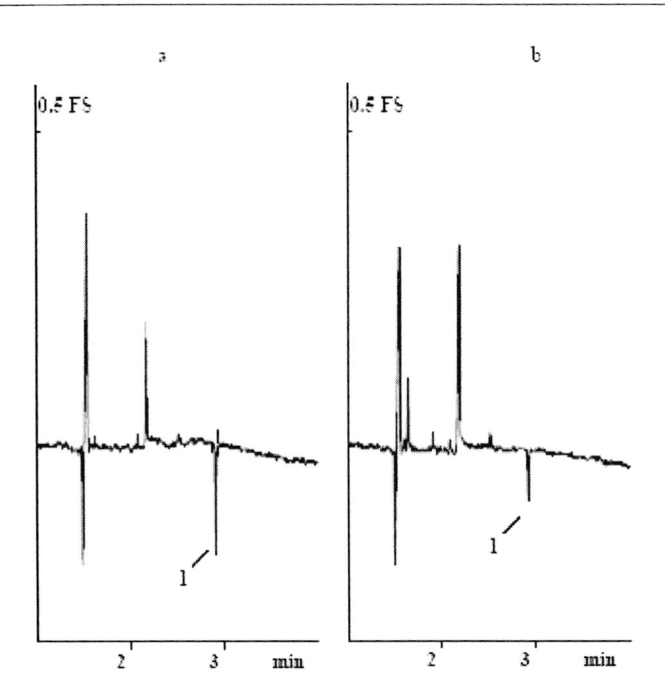

Fig. 4. Electropherogramms of purified fractions after separation on Dowex AG1-X10 resin a) № 23; b) № 25. Dilution was 1010 times. 1= 3-dehydroshikimate

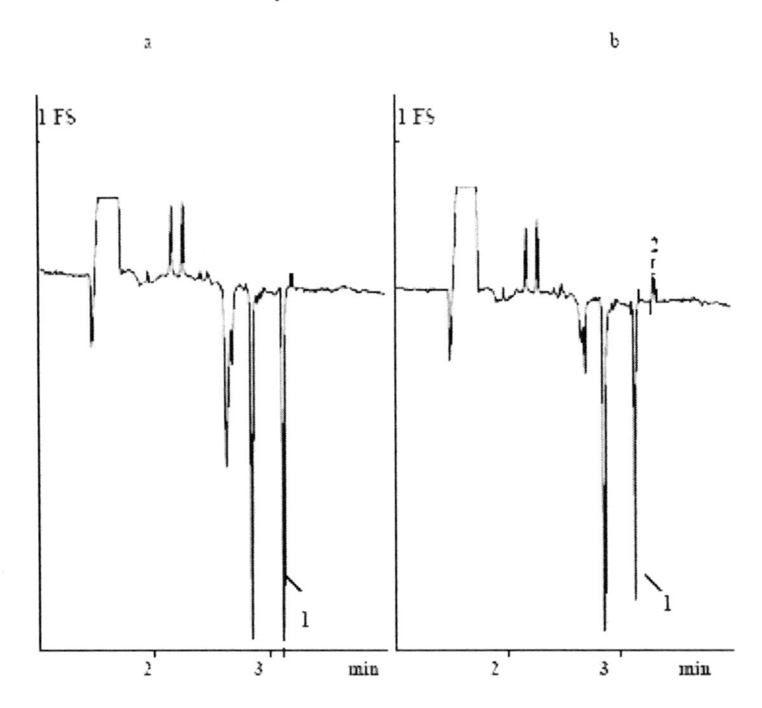

Fig. 5. Electropherogramms of the reaction mixtures, incubated with crude extract, prepared from Bac.subtilis I-118/pDP4 strain, for measuring of direct shikimate dehydrogenase activity during: a) 0 min; b) 20 min. Dilution was 20 times. 1= 3-dehydroshikimate; 2= shikimate

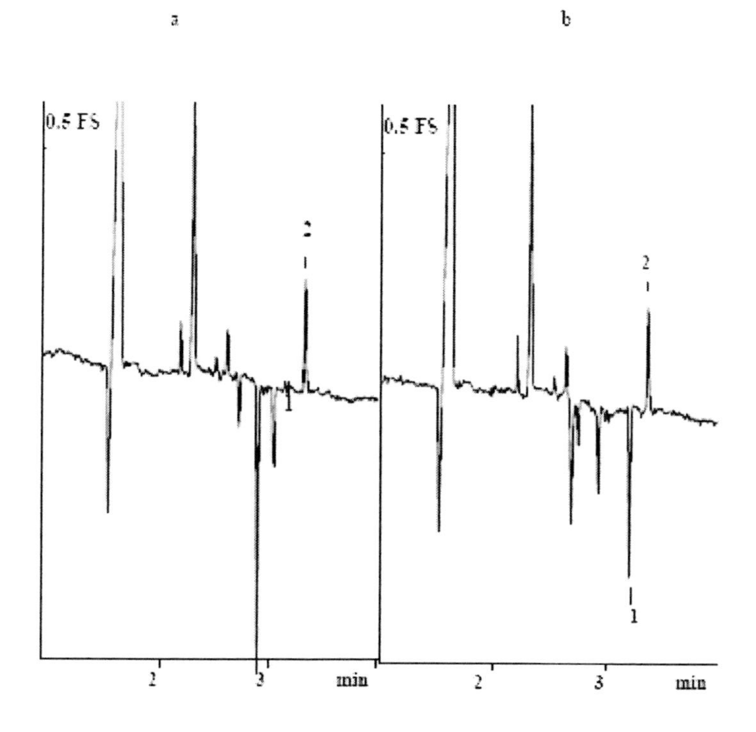

Fig. 6. Electropherogramms of the reaction mixtures, incubated with crude extract, prepared from Bac.subtilis I-118/pDP4 strain, for measuring of shikimate dehydrogenase activity in reverse direction during: a) 0 min; b) 10 min. Dilution was 40 times. 1= 3-dehydroshikimate; 2= shikimate

Table III. Activity of shikimate dehydrogenase in crude cell-free extracts of different strains

Strain	Specific activity nM/min*mg
B-1403	0
I-118	6.6
B-1403/pDP4	391.0
I-118/pDP4	382.2

4. Conclusions

The use of different methods for the complex investigation of samples gave the possibility to make the full identification of 3-dehydroshikimic acid, as an important indicater of biosynthetic disorders. CE was used for determination of shikimic acid and dehydroshikimic acid in fermentation broths of various mutant strains. Besides the CE method presented here is suitable for identification and analysis of these compounds in reaction mixtures for measuring of shikimate dehydrogenase enzyme activities.

ACKNOWLEDGEMENT

We are very grateful to *Dr. Jomantas J. (CSC AGRI)* for gifts of Bacillus subtilis strains: I-118, I-118/pDP4 and I-118/pBAG7.

REFERENCES

[1] R.C. Millican: *Methods Enzymol.*, 17A, 352 (1970).
[2] W.A. Pline, J.W. Wilcut, S.O. Duke, K.L. Edmisten, R.Wells: *J. Agric Food Chem.*, 50(3), 506 (2002).
[3] D.M. Mousdale, J.R. Coggins: *J. Chromatogr.*, 329(1), 268 (1985).
[4] M. Kordiš-Krapež, V. Abram, H. Kac, S. Ferjančič: *Food technol. biotechnol.*, 39(1), 93 (2001) .
[5] B.M. Silva, P.B. Andrade, G.C. Mendes, R.M. Seabra, M.A. Ferreira: *J. Agric Food* Chem., 50(8), 2313(2002).
[6] H. Yamamoto, K. Yazaki, K. Inouc: *J. Chromatogr. B Biomed. Sci. Appl.*, 738(1), 3 (2000).
[7] E. Gollub, H. Zalkin, D.B. Sprinson: *Methods Enzymol.*, 17A, 349 (1970).
[8] U.S. Maitra, D.B. Sprinson: *J. Biol. Chem.*, 253(15), 5426 (1978).
[9] S. Mitsuashi, B.D. Davis: Biochim. *Biophys. Acta*, 15(1), 54 (1954).
[10] D. Balinsky, A.W. Dennis: *Methods Enzymol.*, 17A, 354 (1970).
[11] J.R. Coggins, M.R.Boocock, S. Chanduri, J.M. Lambert, J. Lumsden, G.A. Nimmio, D.D. Smith: *Methods Enzymol.*, 142, 325 (1987).
[12] O. Adachi, S. Tanasupawat, N. Yoshihara, H.Toyama, K. Matsushita: *Biosci.Biotechnol.Biochem.*, 67(10), 2124 (2003).
[13] H. Engelhardt et al.(Eds.): Guide to Capillary Electrophoresis. *Mir*, Moscow, 1996. 232p. (in Russian).
[14] D. N. Heiger: *High Performance Capillary Electrophoresis.* Hewlett Packard Company. Publication Number 12-5091-6199E, 1997. 140p.
[15] R. Kuhn, S. Hoffstetter-Kuhn: *Capillary Electrophoresis: Principles and Practice.* Springer laboratory, 1993. 378p.
[16] F.K. Beverly: *J. Chromatogr.*, 546(1), 423 (1991).
[17] A.E. Novikova, L.V. Ivanovskaya, M.M. Gusyatiner: *Biotechnology*, 2(1), 89 (2004). (in Russian).
[18] M. Willetts, P. Clarkson, M. Cooke: *Chromatographia*, 43(11), 671(1996).
[19] S.Yuku: *J. Genetics*, 50, 155 (1975).
[20] Patent US6436664.
[21] C.H. Duncan, G.A. Wilson, F.E. Yong: *Gene*, 1(2), 153 (1977).
[22] K. Kiritani, R.P. Wagner: *Methods Enzymol.*, 17A, 755, (1970).
[23] A.P. Leech, R. James, J.R. Coggins, C. Kleanthous: *J. Biol. Chem.*, 270(43), 25827 (1995).
[24] S. Chaudhuri, J.M. Lambert, L.A. Mccoll, J.R. Coggins: *Biochem J.*, 239(3), 699 (1986).

In: Industrial Application of Biotechnology
Editors: I. A. Krylov and G. E. Zaikov, pp. 29-37 ISBN 1-60021-039-2

Chapter 4

ELABORATING BASES OF TECHNOLOGY OF ISOLATION OF AMYLASE, LIPASE AND A COMPLEX OF PROTEASES FROM CATTLE PANCREAS

E. A. Dudnikova, A. A. Krasnoshtanova, M. M. Baurina and I. A. Krylov*
Department of Biotechnology,
D.I. Mendeleyev University of Chemical Technology of Russia; Moscow

ABSTRACT

The technological scheme of production of the enzymes such as amylase, lipase and a complex of proteases from a cattle pancreas has been elaborated in conditions of its complex processing with isolation of ribonuclease. It has been established, that isolation of enzymes is carried out according to the uniform technological scheme including stages of extraction, ultrafiltrate concentration and precipitation from water-salt and water-alcohol solutions. The optimum conditions of the given process realization have been chosen.

Key words: bovine pancreas, lipase, amylase, protease, extraction.

Lipases are of great interest for many branches of national economy where partial or full hydrolysis of fats and oils is necessary. They may be applied in food and light industry, agriculture, medicine, in household chemistry, municipal services and in analytical practice [1]. Lipases of an animal origin and less often than a microbic one are used actively in medicine as therapeutic preparations in medical diagnostics [2]. The big hopes are assigned by experts to use lipases in the cosmetic industry, and also at manufacture of furs and leather. The use of lipases in these branches allows deleting lipids easily, giving elasticity and softness to leather, and natural shine and structure of high quality to furs. Elaboration of

* Department of Biotechnology, D.I. Mendeleyev University of Chemical Technology of Russia; 9, Miusskaya sq., Moscow, 125047Russia; fax: (095) 978-74-92; E-mail: krylov@muctr.edu.ru

thermo stable and alkali-proof preparations for use in structure of washing-up liquids is a very actual one in the long term outlook. The prospect of lipase use is important for sewage treatment from fats, at processing household waste products and especial for clearing sewer systems. It is possible to use lipases also successfully in agriculture at preparation of easy-to-digest forages and for improvement of metabolism at animals.

Amylolytic preparations are widely issued in the world. This is basically a large-scale manufacture. Amylases make usage almost in all areas where it is processed starch containing raw material. Amylases are used for saccharification of grain and potato starch. The biggest consumer of amylolytic enzymes is alcohol and brewing trade industries where now malt (the germinate grain) is successfully replaced by amylolytic fermental preparations [1]. These preparations are used in breadmaking, and also in starch treacle manufacture for production of various kinds of treacles, glucose and glucose-fructose syrups. Amylases are used for improvement of quality of concentrates and fast cooked dishes, amylolytic fermental preparations are of big prospect for the industry making washing-up liquids. There thermostable and alkali-proof amylases may be the fine biological additives for removal of carbohydrate pollutions. Amylases are used for desizing fabrics in textile industry and preparation of strong pastes of starch during dyeing fabrics. Recently the use of amylases is paid attention at processing various starch containing wastes into fodder protein preparations. Purified amilases are applied for the analytical purposes and in medicine, and also for manufacture of cyclodextrins [2].

Proteinases are applied in food technology where there is a process with use of microorganisms (yeast, lactic bacteria etc.) [1]. Introduction of proteinases in process and as a result of hydrolysis of fibers of processable raw material allow providing the yeast with normal conditions of ability to live that improves all technological process especially in brewing, alcohol industries, and winemaking [3].

The complex fermental preparations containing proteinases are used in food concentrate and canning industry at preparation of concentrates from hard cooking groats, peas, string beans etc.

Proteinases may be used in the tanning industry for processing leather during its depilating and bating with a high effect: quality of leather is improved, thickness of the ready leather is kept, the separated bristle may be used as secondary raw material, and the main thing - working conditions are sharply improved. Proteinases are used for removal of fiber from the surface of silk thread at processing natural silk [4].

The biggest necessity in proteolytic enzymes is connected to their use in composition of synthetic washing-up liquids. Processing hospital linen polluted with blood and others discharge of protein nature by proteinase containing synthetic washing-up liquids is especially effective.

Proteolytic preparations especially of the animal origin are widely used in medical industry and medicine. They are applied to prepare nutritious and diagnostic mediums, for manufacturing some medical wheys and vaccines. Proteinases of a various degree of purification are used as medical products for regulation of blood curtailing processes on treating for inflammatory processes, for completion of lack of enzymes in an organism etc.

It is necessary for manufacture of all enzymes above mentioned to find a cheap and accessible enzyme containing source. A cattle pancreas has a number of advantages as it contains all necessary components. Therefore the purpose of the given work was to elaborate the scheme on isolation of all these enzymes as now this gland is processed with loss of the important components.

MATERIALS AND METHODS

Frozen biomass of bovine pancreas, containing 25 % of dry substances and having amylase, lipase and protease activities as large as 27 un/g, 0.88 un/g, and 10000 un/g respectively was used as the object of the research.

Research of quantitative regularities of the enzyme extraction from biomass of bovine pancreas was carried out in a glass reactor with the volume up to 1000 ml provided with a mixer and several necks for sampling and also supplied with the electric drive, the thermometer, a water jacket, and electrodes of pH-meter.

At realization of experiences beforehand prepared suspension of a biomass of the crushed pancreas was brought in the reactor heated up to, the temperature required and pH of medium being established. The moment of the ending of heating suspension (the time of heating did not exceed 5 minutes) was accepted as the beginning of extraction process. Through the certain time intervals the samples of the extract were taken for tests. The samples were cooled up to temperature 0-4 ° at once in an ice bath and depending on the purpose of the experiment the activity of amylase, lipase or protease were determined by techniques [5].

For removal of the slime of cattle pancreas after extraction the centrifugation was carried out at 6000 rpm within 10 minutes.

For realization of ultrafiltration concentration of cell-free extracts the hollow fibre devices VPU-15 HA holding substances with molecular weights of higher than 15 кDa were used.

All national reagents which provided realization of analyses were prepared with application of reactants of higher quality.

Evaluation of the experimental data was carried out with the use of standard programs of integration of the differential equations and optimization of their parameters followed by comparison of the results of calculation with the experimental data according to Fisher's test, being used for check of adequacy of the developed mathematical model ($F_{opt} < F_{tab}$).

RESULTS AND DISCUSSION

Complex processing of biomass of cattle pancreas includes extraction from it three groups of enzymes: amylase, lipase and protease.

Amylase is known to be most full extracted by processing cattle pancreas with water-salt solutions, lipase - with organic solvents, proteases - with acid-water solutions. At the first stage of researches it was necessary to establish a sequence of processing of biomass of cattle pancreas by above-stated solutions with the purpose of the greatest possible extraction of one of enzymes at the minimal losses of the others. In result the following sequence of extraction of enzymes was established: 1- amylase, 2- lipase, and 3- protease. The typical curves of extraction of amylase, lipase and protease by appropriate solutions with various initial contents of biomass of cattle pancreas in the medium are given on Fig. 1 (a- c).

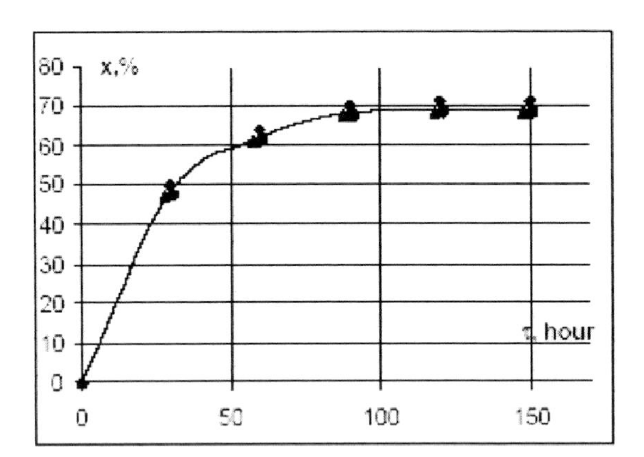

Fig. 1- a. Dependence of a degree of extraction of lipase from concentration of cells insuspension of a pancreas

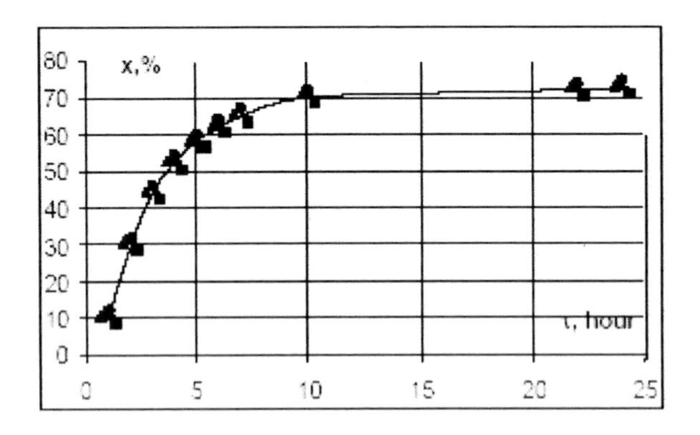

Fig. 1-b. Dependence of a degree of extraction of amylase from concentration of cells in suspension of a pancreas

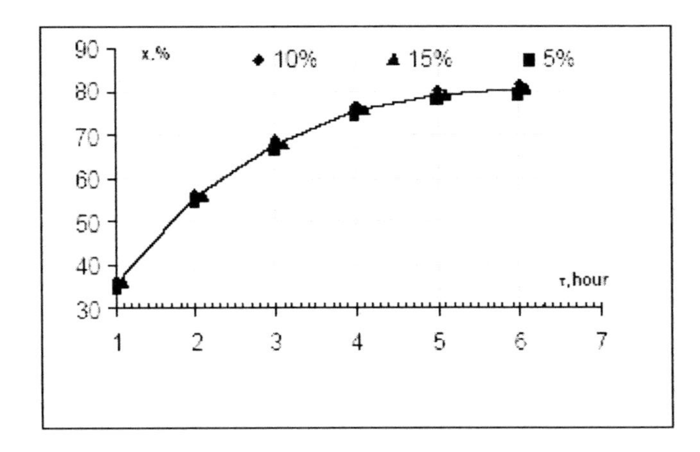

Fig. 1-c. Dependence of a degree of extraction of protease from concentration of cells in suspension of a pancreas

From the submitted data it is visible, that in all cases the degree of extraction of enzymes does not depend on the initial contents of a biomass in suspension; all curves of extraction either have extremum, or achieve quasi-stationary state; on curves of extraction of amylase and protease there is no point of inflection whereas the curve of extraction of lipase has a strongly expressed S-shaped character. All this has allowed assuming the following sheme of consistently - paralleling transformations for the description of extraction process:

For amylase and protease

$$Ecell \xrightarrow{k_1} Es \xrightarrow{k_2} Einact \qquad (1)$$

for lipase

$$Ecell \xrightarrow{k_0} I \xrightarrow{k_1} Es \xrightarrow{k_2} Einact \qquad (2)$$

where E_{cell}, I, E_s, E_{inact} - concentrations of enzymes in cells of cattle pancreas, intermediate, active enzyme in a solution and inactive enzyme, accordingly, k_0, k_1, k_2 - effective constants of rate of the appropriate stages.

Processing experimental data was started with a series of the experiments with the constant temperature, concentration extractive solvent and biomass of cattle pancreas in the medium. At the first stage of researches differential processing experimental data was carried out on initial rates in coordinates ln (1-x) - f (τ), where x - a degree of extraction of enzyme, τ - time. The results of processing are submitted on Fig. 2 from which it is visible, that they represent linear dependences with factor of correlation $r_к > 0.95$. On a tangent of a corner of an inclination of the given dependence to the abscissa axis the values of effective constants k_1 (for amylase and protease) and k_0 (for lipase) have been determined. Other values of constants were established during the subsequent integrated processing experimental data, also optimization of all effective constants found being fulfilled.

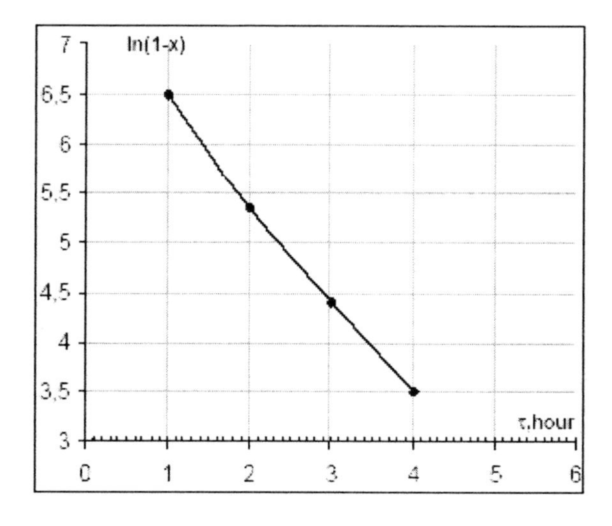

Fig. 2. Typical dependence for differential data processing

At the following a stage of researches we have tried to establish the influence of concentration of extractive solvent and temperatures on the process of extraction.

At researching the influence of temperature on process of enzyme extraction the appropriate series of experimental data which were received by above described approaches were processed. Arrenius equation was used for estimation of the dependency of the numerical values of constants from temperature. As the results of calculations have shown, the function in coordinates ln k_{ef} - f (T^{-1}) where T – the absolute temperature is linear with coefficient of correlation not less than 0.94 (Fig. 3). It has allowed calculating appropriate values of the activation energy and preexponential factor for each sheme of transformations (1) and (2). Results of calculations are given below in table 1.

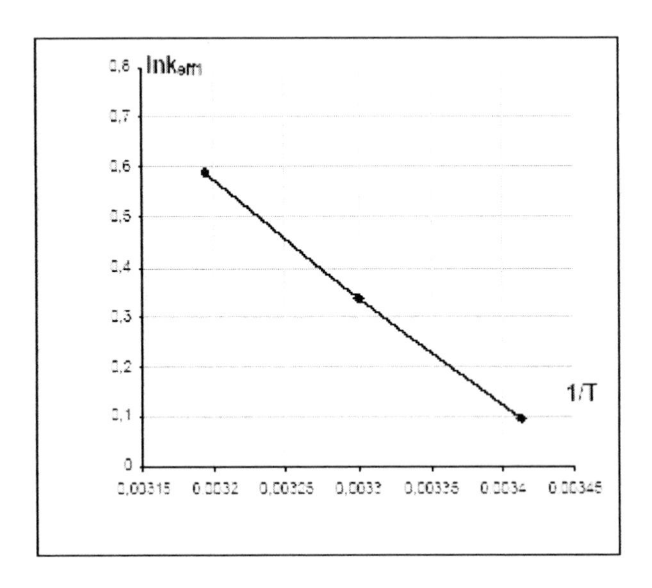

Fig. 3. Typical dependence for processing experimental data on Arrenius equation

Table 1. Values of effective constants of process of extraction of enzymes from biomass of a pancreas

	Protease		Amylase		Lipase	
	k_{eff1}	k_{eff2}	k_{eff1}	k_{eff2}	k_{eff1}	k_{eff2}
E	19±0.9	4±0.4	31±3.0	9±0.9	26±2.5	7±0.7
k_0,h $^{-1}$	0.8475±0.085	0.1795±0.02	0.092±0.01	0.0575±0.005	49.8±5.2	0.0256±0.002
Ka, mol/l	15.07±1.51	10.08±1.01	-	-	-	-
A_0	2.01±0.20	2.204±0.23	$9.2\cdot10^4$±9195	41350±4135	0.2±0.02	1.43±0.15

At studying influence of extractive solvent on process of extraction in the case amylase we have investigated the influence of concentration of sodium chloride

$$k_{eff} = k_o \cdot C_{NaCl} \quad (3)$$

in the case of protease-concentration acid extractive solvent (owing to low concentrations of an acid the parameter of activity of ions H^+ - a_{H+} were used)

$$k_{eff} = k_o / (1+K_a/a_{H^+}) \tag{4}$$

and in a case lipase - electric moment of a dipole (M) (such organic extractive solvents as: ethanol, tret-butanolt, toluene, ethyl acetate, acetone were used for the estimation of this influence)

$$k_{eff} = M \cdot k_o \tag{5}$$

The results of processing experimental data in the coordinates appropriate to every case are submitted on Fig. 4 (a - c).

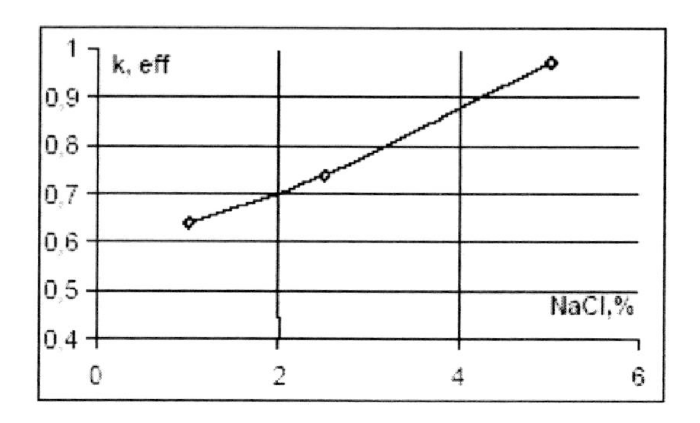

Fig. 4 – a. Typical dependence for calculation of parameters of relationship of effective constant of extraction of lipase from concentration of salt

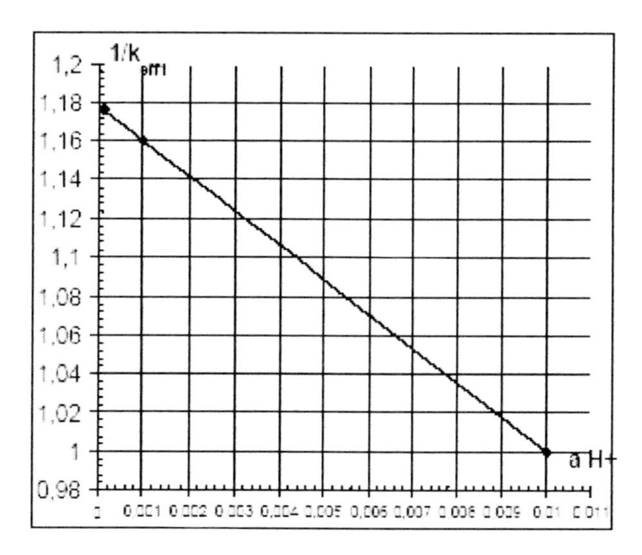

Fig. 4-b. Typical dependence for calculation of parameters of relationship of effective constant of extraction of protease from acidity of the medium

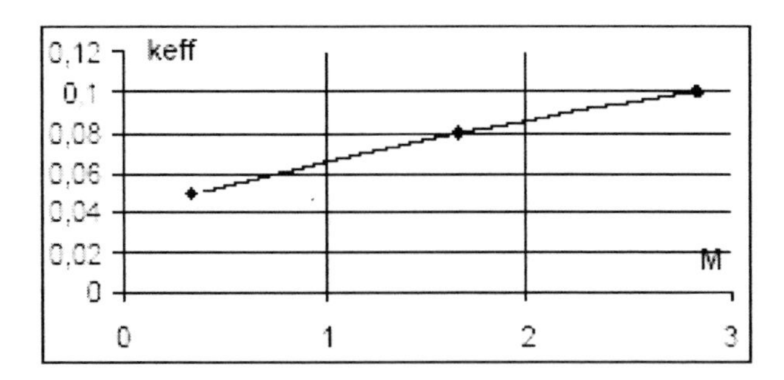

Fig. 4-c. Typical dependence for calculation of parameters of relationship of effective constant of extraction of from amylase from the electric module of extractant

Thus, the shemes of transformations (1) and (2), the equations (3), (4), (5) and numerical values of their parameters have determined mathematical models of processes of extraction adequately describing the experimental data.

We used the received mathematical models for selection of optimum conditions of process of each enzyme extraction. The results of calculations confirmed with experiments have shown that optimum conditions of extraction for amylase are the temperature - 30 °C, concentration of sodium chloride - 5 %, the time of extraction - 4 hours, for lipase - 50 % solution of ethanol, the temperature 20 °C, the time of extraction - 6 hours, and for protease – the temperature 30 °C, pH 2.0, the time of extraction - 6 hours. In all cases the concentration of cells of cattle pancreas in suspension - 10 % and losses of activity in slime of cattle pancreas in each stage do not exceed 20 %.

As the enzymes are high-molecular compounds the most acceptable way of their isolation from cell-free extracts is ultrafiltration concentration. Ultrafiltration was carried out at the hollow fibres VPU-15 HA which detain substances with molecular weights higher than 15 кDa. For each case the parameters of ultra filtrational installation were also estimated, such as integrated selectivity, specific productivity and activity of enzyme, corresponding to gelatinization concentration. The results of the appropriate calculations are shown in table 2.

Table 2. Parameters of process of ultrafiltration concentration of cell-free enzyme containing extracts

The extract containing	Specific productivity, G, l/m^2h	Integrated selectivity, φ, %	activity of enzyme, corresponding to gelatinization, Ag, units/ml	Concentration degree, n
Amylase	1,2	80	138	5
Lipase	1,2	85	69	8
Protease	1,2	85	7880	6

From the data submitted in table 2 it is visible, that the use of the given type of ultrafiltration is effective, as the values of selectivity amount to 80-85 % that allows increasing the activity of concentrated products received and increasing their degree of precipitation.

The following stage of researches became studying a stage of precipitation of fermental preparations from concentrates.

This problem is solved most simply in a case of lipase as for its precipitation it is simple enough to cool concentrates up to the temperature 4 - 8 °C that will result in precipitation of lipase not less than 90 %. In a case of amylase and protease the influence of pH of the medium and the hydro module of alcohol was investigated. The following data were obtained: for amylase: the optimum range of pH 3.0 – 4.0 and the relation of amylase: isopropanol 1:8; for protease: an optimum range of pH 2.5 – 3.5 and the relation of protease: ethanol 1:10.

REFERENCES

[1] I.I. Grachyeva, A.Y. Krivova: *Technology of enzyme preparations*, Moscow, 2000 (in Russia).

[2] E.P. Yakovenko: Enzyme preparations in clinical practice. *Clinical pharmacology and therapy*, 1998, №1. 17-20 p (in Russia).

[3] L.V. Antipova: *Application of enzyme preparations of proteolitic activity in brewing-nonalcocholic industry*, Moscow, 1992 (in Russia).

[4] L.V. Antipova: *Biotechnologycal aspects of rational use of secondary raw material of meat processing industry*, Moscow, 1991 (in Russia).

[5] G.V Polygina., V.S.Cherednichenko, L.V. Rimaryeva: *Definition of activity of enzymes*, Moscow, 2003 (in Russia).

In: Industrial Application of Biotechnology
Editors: I. A. Krylov and G. E. Zaikov, pp. 39-44
ISBN 1-60021-039-2

Chapter 5

Aseptic Cultivation of Sowing Culture of Alcohol Yeast in Bioreactors with Membrane Aerators

V. M. Emelyanov*, J. P. Aleksandrovskaja and S. G. Mukhachev
The Kazan State Technological University, Kazan

Abstract

The technological unit for cultivation of pure alcohol yeast culture in aerobic conditions with guaranteed maintenance of aseptic conditions is offered. The main part of the unit is a bioreactor with inserted cartridge with tube type membranes made of nonporous polymeric material. The membranes of solubility are selective in relation to components of a gas mixture. In order to increase the economic characteristics of the unit, the block of preliminary enrichment of air with oxygen on the basis of use of liquid membranes is installed in its structure.

Key words: alcohol yeast, pure culture, aerobic cultivation, aseptic, nonporous membranes, enrichment of air.

One of the ways to decrease infection in alcohol manufacture and to increase yeast activity is the cultivation of pure culture in sterile conditions by intensive aerobic technology [1, 2].

The realization of effective aerobic processes of yeast cultivation is attended with a problem of maintenance of conditions of oxygen delivery and low rate of oxygen extraction from bubbled gas. The exception of infection via aeration and mixing systems will allow to solve aseptic problem in fermenters and, thus, to increase technical and economic parameters of alcohol manufacture.

The new approach to a problem of aeration in process of aerobic cultivation of sowing materials in alcohol manufacture is offered, it consists in delivery of pure oxygen or air,

* The Kazan state technological university. 420015, Kazan, K. Marx street, 68. E-mail: biotex_kgtu@rambler.ru; Fax: 8432-367542

enriched with oxygen, to a bioreactor through nonporous homogeneous polymer tubular membranes. The achievement of completely aseptic conditions is provided because gaseous components are reliably sterilized while passing through a nonporous membrane.

The main part of the unit is a bioreactor with inserted aerating device on the basis of a tubular membrane. The aeration is carried out through the walls polymer tube, in which the oxygen or air, enriched with oxygen, is fed at excessive pressure. During this process the deformation of a membrane occurs, its surface increases, and the wall becomes thinner. The variation of pressure can be executed within the limits of strength of a membrane.

Only the membranes of solubility, selectively penetrative for oxygen molecules, can be used for oxygen introduction without bubbles into cultivated liquid. Polydimethylsiloxane was used as a membrane material for this study.

Transport of oxygen through tubular homogeneous silicon membrane was experimentally investigated. The results are given in Table 1.

Transfer of oxygen through a membrane in conditions of chemical reaction was previously measured. Sulfite technique was used for a quantitative estimation of rate of adsorption of oxygen. Mass exchange characteristics of bioreactor at a variation of excessive pressure of oxygen in a tubular membrane were studied. The results have demonstrated that sulfite numbers and rate of transport of oxygen through a membrane practically coincide in the whole working range of pressure change. It shows that all oxygen is consumed at once, without bubble formation.

Table 1. Specific rate of the oxygen diffusion through a siliconous membrane at a variation of excessive pressure in a cavity of a membrane

Pressure of oxygen, atm.	Area of a surface of a membrane, cm^2	Rate of the oxygen diffusion through a membrane, gO_2/Hour	Specific rate of the oxygen diffusion through a membrane, $gO_2/Hour \cdot m^2$
0,5	186,9	0,0365	1,95
1,0	192,5	0,0851	4,42
1,5	197,2	0,1237	6,27
2,0	206,5	0,1697	8,22
2,5	356,3	0,2317	6,50
3,0	390,0	0,3027	7,59
3,5	464,8	0,4791	10,31
4,0	522,8	0,5702	10,91

In order to study the process of alcohol yeast cultivation in aerobic conditions in membrane bioreactor the following was carried out:

- Series of experiments on periodic cultivation of alcohol yeast culture of Saccharomyces cerevisiae XII races with fractional feeding with substrate;
- Subtracting-filling up cultivation of yeast;
- Continuous cultivation of yeast.

Glucose was used as a substrate. Rieder modified medium was used as nutrition.

Temperature and level of acidity of cultural liquid, difference of pressure of oxygen on a membrane, presence of extraneous microflora were controlled, concentration of reducing substances in cultural liquid by Berthran method, quantity of cells by means of the Goryaev chamber and concentration of alcohol yeast by method of optical density were determined during the experiments.

The laboratory automatic fermentation unit was used in the experiments, which includes:

- Bioreactor with operation volume of 1 litre and membrane aerating device;
- Control block of technological parameters of fermentation process;
- Technological supplement and auxiliary equipment for feeding nutritious media and oxygen in bioreactor.

Reactor is supplied with three-level turbine stirrer with face sealing. The range of speed of stirrer rotation is 120-1200 turns per minute. Hydrodynamic principle of destruction of foam by transferring of recirculation flow of fermentation media on a rotating disk of foam quenching unit was used for quenching of foam without using of chemical substances. The aeration is carried out through the walls of silicone hose with closed end fixed on a cylindrical body made of corrosion-proof steel, established in the reactor. Length of silicone hose is 182 cm, external diameter is 0,38 cm, thickness of a wall is 0,1 cm. Reactor is equipped with connecting pipes for feeding components of nutritious media and for sampling. The gauges for measurement of physical and chemical parameters are established in measuring circulating cuvette, through which fermentation medium is continuously pumped.

Fermenter together with it supplement and vessels for liquid components were sterilized in autoclave during 1,5 hours at excessive pressure 0,1 Mpa, providing complete absence of cases of culture infection. Technological equipment was assembled in a flame of alcohol torch. Ph-electrodes were sterilized by hydroden peroxide g 20 minutes, and pO2-electrode was sterilized in 70% solution of ethyl alcohol during 2 hours before installation in the reactor.

Pure culture of alcohol yeast Saccharomyces cerevisiae XII races grown on agar, was put in 750 мл shaking vessels containing 100 ml sterile nutritious medium, which is sterilized in medical autoclave at 0,05 Mpa during 30 minutes and cooled down to 30°C. The culture was shaking in shaking vessels on a shaker with shaking rate of 220 min^{-1}, inserted in oven, at temperature 30°C during 10-12 hours. The grown culture was put under a flame of alcohol torch into membrane bioreactor with operation volume 0,5 litre on sterile nutritious medium. Technical oxygen was supplied to the bioreactor from a cylinder through cylindrical silicone membrane. We were raising pressure of oxygen in a membrane cavity from 1 to 4 atm as the concentration of biomass was increasing.

The duration of periodic cultivation of alcohol yeast in membrane bioreactor was 11-12 hours. During the experiment samples for the analysis were taken immediately after inoculation (zero sample) and after the 2 hours expiration in each hour of growth. order supervision over a Condition of yeast cells was monitored under microscope, quantity of cells was calculated in the Goryaev chamber, mass of yeast suspension on photocolorimeter (КФК-2-УХЛ 4.2) was determined during growth of culture. Concentration of reducing substances in cultural liquid was determined by Berthran method. Feeding with 50% glucose syrup was making during cultivation when substrate was coming to an end. Temperature and pH of the media were maintained on an optimum level with the help of control devices of 2 class. At the end of the process presence of extraneous microflora was determined by inoculation of cultural liquid sample on a solid nutritious media. Foam forming was controlled visually.

As the control variant anaerobic cultivation was carried out with other equal conditions.

The results of periodic process are represented in Table 2.

Table 2. Concentration of absolutely dried biomass, cultivated in aerobic and anaerobic processes at various density of sewage

Number of experiment	Concentration of absolutely dried biomass, g/l		Increase of absolutely dried biomass in experiment relatively to the control, %
	experiment	control	
1	11,4	5,3	115
2	18,9	5,8	226
3	23,5	7,1	231

The scattering of the data of relative gain of biomass concentration given in Table 2 is caused by various density of inoculation. In all cases the essential difference between intensity of growth process in conditions of experiment and control experiment was obtained. Intensification of growth process in aerobic conditions was not accompanied by appreciable change in foam forming.

As an example of process realization the dynamics of change of the basic parameters in experiment № 2 is shown in a Fig. 1.

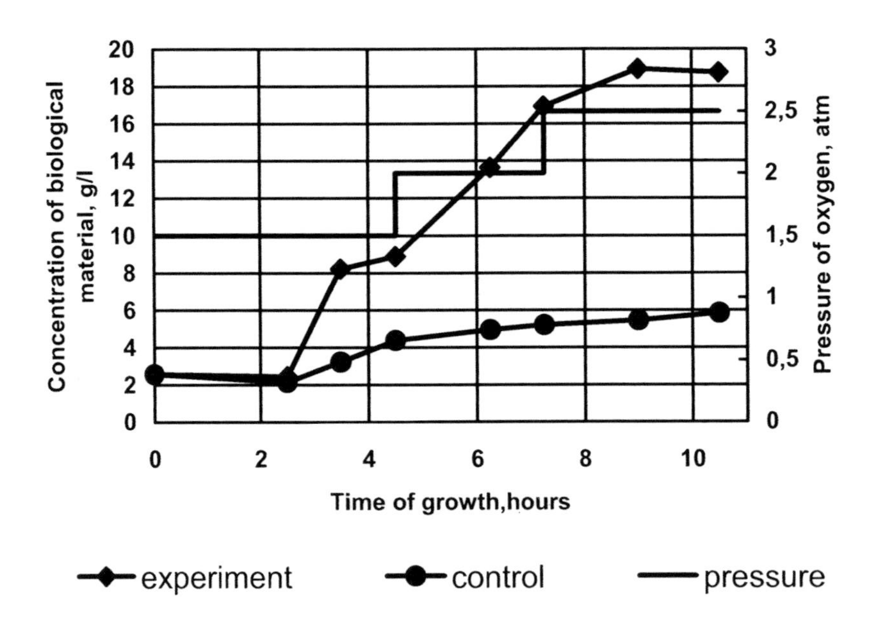

Fig. 1. Growth of biomass in aerobic and anaerobic periodical processes

As a result of realization of subtracting-filling up process in aerobic and anaerobic conditions a difference in a gain of biomass is also obtained. In anaerobic process biomass has grown up to 4,62 g/l, and in aerobic - up to 7,21 g/l after 4 hours of cultivation, that is 1,5 times higher. In comparable conditions during 9,25 hours of cultivation biomass of yeast (accounting to absolutely dried biomass, ADB) in experiment has grown from 1,8 to 10,6 gADB/l, in the control experiment - from 1,8 to 6,2 gADB/l. The average rate of yield of biomass in aerobic process was 2,8 g/l·hour, whereas in anaerobic - 1,7 g/l·hour.

Continuous cultivation was carried out after attainment in periodic fermentation of the end of exponential phase and beginning of a stationary phase of process. A series of experiments with D from 0,1 to 0,28 $hour^{-1}$ was carried out. The characteristics of continuous process: speed of flowing, concentration of biomass, concentration of residual sugars, concentration of the dissolved oxygen were registered in the established condition. Thus the average values during last 5 hours were taken as parameters of stationary process.

The maximal rate of yield of biomass was observed at the rate of dilution of 0,23 $hour^{-1}$ and was 2,89 gADB/l. It is necessary to note, that the increase of specific flow speed from 0,1 to 0,17 $hour^{-1}$ does not cause essential growth of residual sugars concentration. The further increase of rate of dilution results in significant growth of concentration of residual sugars from 1 g/l at D=0,18 $hour^{-1}$ to 11 g/l at D=0,2 $hour^{-1}$, that indicates low of oxygen in this area. The rate of mass transfer of oxygen can be increased either by increasing the area of a membrane surface or by increasing oxygen pressure in its cavity. Since of residual sugars can be utilized at the subsequent stages in alcohol manufacture, it is reasonable to carry out the fermentation process at the rate of dilution of 0,2-0,25 $hour^{-1}$. In this case productivity of fermentation process will be entirely defined by the area of membrane surface and by difference of oxygen pressure on a membrane.

So the realized processes have shown increase of yield of alcohol yeast biomass in 2-3,3 times compared with anaerobic processes which have been carried out in similar conditions.

The processes were carried out in aseptic conditions at all modes of yeast cultivation.

All carried out processes were characterized by simplicity of control of oxygen supply.

The investigated cultivation processes are easily scaled by preservation of a specific surface of a membrane and can be realized in industrial scales.

In order to reduce the expenses we offered to use air, enriched with oxygen, instead of pure oxygen, in yeast cultivation in membrane bioreactor. The enrichment of air with oxygen can be carried out by the use of liquid passive perfluorocarbon membranes in bubbling column with inserted contact appliances with natural or compulsory circulation, for example, in bubbling column with Koch appliance.

The oxygen is better dissolved in perfluorodecaline in comparison with nitrogen. The opportunity of it transfer in technological process with the help of pumps allows to design continuous process of air separation.

The calculation of system of air enrichment by oxygen on passive perfluorocarbon membranes, executed with the help of the simulating program ChemCad, has shown an opportunity of obtaining of air mixture containing up to 77% oxygen, after three steps of enrichment, each of which is constructed according to the same scheme [3], given on Fig. 2.

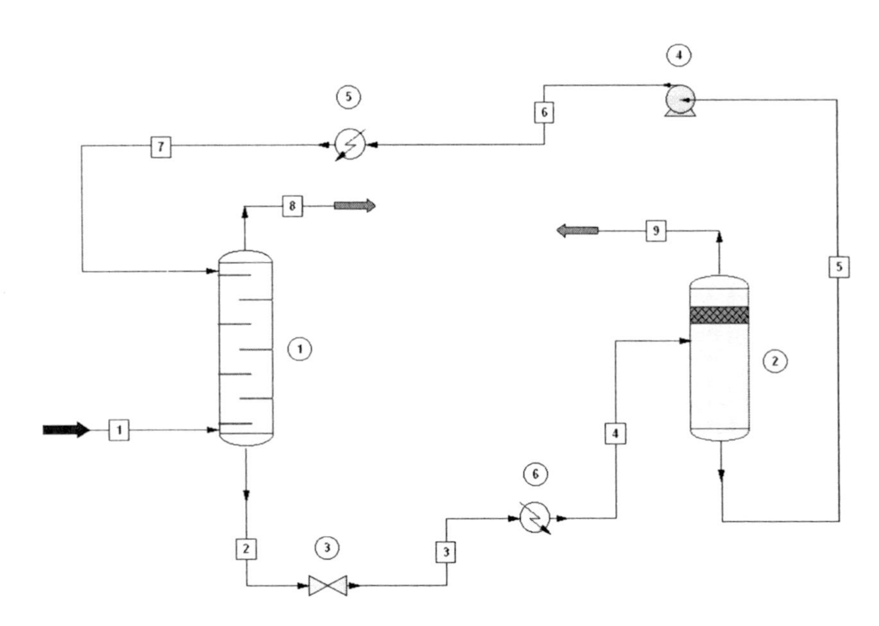

1 – absorber; 2 – desorber; 3 – valve; 4 – pump; 5, 6 – heat exchanger

Fig. 2. Technological scheme of the process of air enrichment with oxygen by the employment of liquid perfluorocarbon membranes

The aeration by a mixture containing up to 77% oxygen through silicone membrane will provide supplement of a gas mixture containing about 88% oxygen due to membrane selectivity to the cultural liquid.

Thus, the technological process of aerobic cultivation of alcohol yeast with high aseptic characteristics is offered. It consists of:

- Three step system of air enrichment by oxygen on passive perfluorocarbon membranes allowing to obtain up to 77% oxygen in nitrogen - oxygen mixture;
- System of aeration on a basis of a silicone membrane allowing to realize oxygen supply to the cultural liquid without bubbles and to lower intensity of foam forming.

References

[1] Patent 2136746 Russian Federation.

[2] V.M.Emelyanov, I.S.Vladimirova, J.P.Alexandrovskaya et al.: A biotechnological complex of a division department of pure culture of alcohol yeast: Biotechnology – modern state and prospects of development. *Materials of the international congress.* Moscow, 2002, p 96.(Russian).

[3] J.P.Alexandrovskaya, N.N.Ziatdinov, V.M.Emelyanov et al.: The automated calculation of system of air enrichment by oxygen: Heat and mass transfer processes and apparatus for chemical technology. *The interuniversity thematic collection of the proceedings.* Kazan, 2004, p 84.

In: Industrial Application of Biotechnology
Editors: I. A. Krylov and G. E. Zaikov, pp. 45-53
ISBN 1-60021-039-2

Chapter 6

VORTEX REACTORS FOR HETEROGENEOUS BIOCATALYTICAL PROCESSES

Galina A. Kovalenko*, Sergey V. Sukhunin** and Larisa V. Perminova

*Boreskov Institute of Catalysis, Prosp. Akad. Lavrentieva 5, Novosibirsk 630090, Russia

**Institute of Hydrodynamics, Prosp. Akad. Lavrentieva 11, Novosibirsk 630090, Russia

ABSTRACT

Research and development of novel types of reactors for the heterogeneous biocatalytical processes were carried out. The main directions for the reactor design were the significant intensification of mass transfer of substrate to immobilized enzyme and elimination of stagnant zones. The vortex reactors – *rotor-inertial bioreactor (RIB)* and *vortex-immersed reactor (VIR)* were developed for the diffusion-controlled biocatalytical processes. The lab-scale setups of these reactors were studied in dextrin hydrolysis that was limited by external and internal diffusion of high-molecular weight molecules of dextrin to glucoamylase immobilized on the carbon-containing inorganic supports. It was found, that under the optimal operation conditions, the process productivity and the observed activity of the heterogeneous biocatalysts were an average of 1.2-1.5 times higher in the vortex reactors than those in the fixed-bed reactor.

Keywords: rotor-inertial bioreactor, vortex-immersed reactor, heterogeneous biocatalysts, immobilized glucoamylase, and starch dextrin hydrolysis.

* Fax +7 383 2 30 80 56; E-mail: galina@catalysis.nsk.su

INTRODUCTION

Research and development of novel reactors for biocatalysis and biotechnology permitting a significant increase of process productivity are certainly of importance today. The results of such investigations can be utilized for updating of the existing productions or development of innovative competitive technologies. The design of the reactors calls for joining forces of specialists from different research areas such as biocatalysis, hydrodynamics, enzymology as well as design engineers.

On the development of novel reactor types, significant intensification of mass transfer of a substrate to a heterogeneous biocatalyst is of crucial importance to overcome diffusion resistance of the process. One way to intensify mass transfer is to generate a vortex movement in the liquid. The number of the researches devoted to vortex bioreactors for biocatalysis is not numerous. In ref. [1] the authors came up with an idea to use the Taylor vortex flows in the performance of heterogeneous biocatalytical processes and theoretically justified a possibility of development of such bioreactors. This idea was then embodied in a Taylor-Poiseulle *Vortex Flow Reactor (VFR)* constructed from inner and outer cylinders that were vertically arranged. The outer cylinder was filled with a substrate solution whereas the inner cylinder was immersed into this solution and rotated. The reactor was tested in the enzymatic process of glucose isomerization by immobilized glucose isomerase [3,4]. The heterogeneous process of glucose isomerization was performed also in a vortex type reactor with a tangential flow of two impinging streams [5].

The problem of mass transfer intensification is especially urgent if enzyme utilizes high-molecular weight substrate; for example, glucoamylase catalyzes the exogenous hydrolysis of starch dextrin molecules with Mw. $\geq$ 4000 D. It was found, that the heterogeneous process of dextrin hydrolysis as a whole was limited by diffusion of this substrate to immobilized glucoamylase [6]. Undoubtedly, an indispensable request for the process commercialization is the overcoming of diffusion resistance and the kinetics performance of the process.

This work was aimed at developing novel reactors – rotor-inertial bioreactor (RIB) and vortex-immersed reactor (VIR) for heterogeneous biocatalytical processes controlled by diffusion of a substrate to an enzyme immobilized on the porous support. The reactors were studied for the process of dextrin hydrolysis to starch treacle. The heterogeneous biocatalysts were prepared by adsorptive immobilization of glucoamylase on the macrostructured and granular carbon-containing inorganic supports. The optimal operation conditions of the dextrin hydrolysis were determined for these novel reactors. The efficiency of the process performed in vortex reactors was compared with one in a fixed-bed reactor.

EXPERIMENTAL

Rotor-inertial bioreactor (RIB). The main functional part of the RIB was a container fitted by foam-like heterogeneous biocatalyst (Fig. 1). This cylinder container was rotated around a horizontal axis at the varied rate from 3 to 170 rpm. The total volume of the container was 1.2 liter; the volume of the substrate solution was 0.6 liter; the catalyst volume was 360 cm^3. A thermal unit, consisted of a temperature controller, a heater, a disaster protector and a temperature sensor, permitted one to hold and control a temperature of

50 ± 5^0C. Both heating and feeding of the substrate solution was performed through a channel passing through a thermal unit to a back wall of the container (Fig. 1). A liquid peristaltic pump provided feeding of a substrate through the RIB; the rate of feeding flow was varied from 0.6 to 300 ml/min. A lab-scale setup of the RIB had the following dimensions ~ 20 × 40 × 50 cm [6,7]. The RIB operated with a periodic mode and with a continuous one.

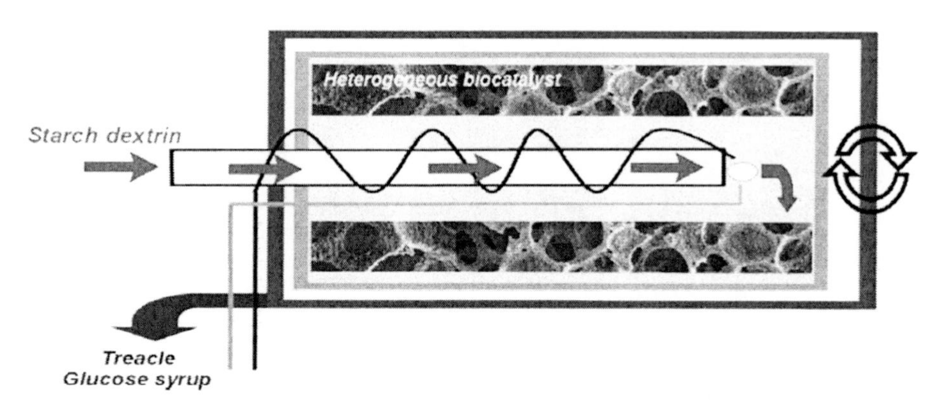

Fig. 1. *Rotor-inertial bioreactor (RIB)* container fitted by the heterogeneous biocatalyst based on foam-like ceramics and immobilized glucoamylase

Vortex-immersed reactor (VIR) was consisted of a rotated confuser-type body filled with granular heterogeneous biocatalysts (Fig. 2) and thermostatic at 50°C reservoir for a substrate solution, in which reactor body was immersed and rotated at varied rate from 50 rpm to 1500 rpm. [8]. The volume of body was 25 cm^3; the volume of substrate solution was 0.5 liter; the biocatalyst volume was 15-20 cm^3. The reactor body was constructed of two parts — an upper and lower scuff plates. The profile of each scuff plate followed the hyperbolic equation $h = 1/R$, i.e. its height h in each point was inversely proportional to radius R. Due to reduced pressure in the center of the twisting liquid, a substrate solution was drawn in 10 mm-hole of the lower scuff plate. As the substrate solution was sucked inside the body, it moved through a biocatalyst bed toward side holes and repeatedly circulated for the reaction time (0.5-1 hour). The rate of substrate circulation through the biocatalyst bed increased with increasing body rotation rate (ω) as $\omega^{1/2}$. During this circulation the dextrin hydrolysis occurred. Biocatalyst granules of 2-4 mm in size were placed into the lower scuff plate of the body; the upper scuff plate was fastened on the top. For small fraction granules of 0.2-1 mm in size, a toroidal nylon filter was made and placed inside the reactor body. The VIR operated with a periodic mode.

For comparison of efficiency of dextrin hydrolysis a well-known *fixed-bed reactor* (*FBR*) was used. This reactor contained a thermostatic at 50^0C glass column 10 cm in length and ca. 1 cm in diameter that was filled with the heterogeneous biocatalysts in study. The height of the granular biocatalyst bed was varied from 0.5 cm to 6 cm. A feeding flow rate (υ) was varied from 3 ml/min to 550 ml/min by liquid peristaltic pump.

Heterogeneous biocatalysts for the process of dextrin hydrolysis have been obtained by adsorptive immobilization of glucoamylase on the carbon-containing supports differing by macrostructure (foams, granules), porosity as well as morphology of the carbon layer synthesized on the surface [9-11].

Thus, the RIB container was fitted by the glucoamylase immobilized on the foam-like macroporous ceramics coated by catalytic filamentous carbon (CFC) [9]; specific surface area was 4 m^2/g. The VIR body was filled with the glucoamylase immobilized on the granular mesoporous carbon support Sibunit and bulk CFC [10,11], specific surface areas of the supports were ca. 440 and ca.160 m^2/g, respectively (Table).

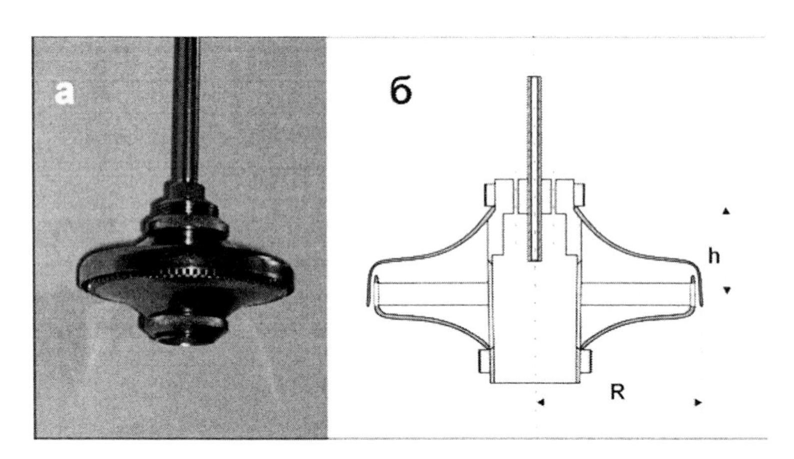

Fig. 2. *Vortex-immersed reactor (VIR)* body he filled with heterogeneous biocatalyst based on granular carbon support and immobilized glucoamylase

Table 1. Biocatalytical activity and stability of glucoamylase immobilized on granular mesoporous carbon supports

Supports	$S_{SP. BET}$, m^2/g	Average pore radius, μ	Adsorption, mg/g		Activity of the biocatalyst *, U/g		Half-life time ($t_{1/2}$), months
			Granules size 0,2-1 mm	Granules size 2-4 mm	Granules size 0,2-1 mm	Granules size 2-4 mm	
Sibunit	441	0,02	23,7	4,2	616	53	11
CFC	162	0,01	4,3	2,0	371	44	10

* – Activity was measured at rotation rate of reactor body equal to 600 rpm

Substrates of the glucoamylase were corn or wheat or potato dextrin prepared as 1–30w/v % buffer solutions (0.05 M acetate buffer, pH 4.6). When reactors operated with a periodic mode, the process efficiency was estimated by the rate of dextrin hydrolysis or the observed biocatalytical activity that were measured and expressed as μmoles of glucose generated per 1 min per 1 g of biocatalyst (U/g). When reactors operated with a continuous mode, the process efficiency was estimated by the dextrin conversion which was determined on the outlet of reactor and expressed (in %) as ratio of amount of glucose generated to total amount of dextrin.

RESULTS AND DISCUSSION

Rotor-inertial bioreactor (RIB). Mass transfer in RIB was intensified due to action of both centrifugal and gravitation forces by following manner. A foam-like biocatalyst into the rotated container (Fig. 1) captured a portion of the substrate solution at the container bottom and moved it upwards. At the container top, these solutions fell down to the bottom due to gravity. The optimal rotation rate of the container was estimated by equation $R\omega^2 = g$, where R was inner radius of the container, meter, ω – angular rotation rate, radian per second, g – gravity acceleration, m/sec^2. This rate was approximately equal to $\sqrt{\frac{g}{R}}$. In this case, the centrifugal force and gravity were balanced at the upper point of the container (liquid fell down from top to bottom). At the lower point of the container, centrifugal force and gravity were summed (liquid moved upwards). The calculation suggested that to provide maximal intensification of mass transfer in the lab-scale setup in study ($R \sim 5$ cm) the optimal rotation rate had to be ca. 140 rpm. As mentioned above, substrate solution was supplied through the inner channel to the back wall of the container. In this case, the solution was additionally curled due to its movement through a foam-like biocatalyst toward the reactor outlet.

Study and optimization of the rotor-inertial bioreactor were performed for the starch dextrin hydrolysis. There were two operation modes for RIB – a periodic mode and a continuous one.

RIB with a periodic operation mode. The effect of rotation rate of the RIB container fitted by the biocatalyst on the dextrin hydrolysis was studied. As follows from Fig. 3, a, the maximal hydrolysis rate was observed if the rotation was higher than 80 rpm. Thus, the diffusion resistance was overcome at this rotation rate. Similar results were observed for corn or wheat or potato dextrin.

This mode of process performance in RIB was compared with a circulation operation mode of the fixed-bed reactor (FBR). A comparison of the process efficiency suggested that the biocatalytical activity under the absence of diffusion resistance, namely, at $\omega \geq 80$ rpm in the RIB (Fig. 3, a) and $\upsilon \geq 50$ ml/min in the FBR (Fig. 3, b) was an average of 1,2-1,3 (sometimes up to 2-3) times higher in the rotor-inertial bioreactor. This observation was probably associated with the presence of "dead" stagnant zones in the fixed-bed reactor filled with foam-like biocatalyst, which was absolutely impossible in the RIB.

RIB with a continuous operation mode. In this case, the container with a biocatalyst was rotated in a substrate-feeding stream flowing through the bioreactor. Studying the effect of container rotation rate on conversion of the corn dextrin, it was shown that the maximal conversion was also observed at ≥80 rpm (Fig. 4). Note that there was more pronounced maximum at the curve as compared to the periodic operation mode of the RIB (see Fig. 3).

On studying the effect of feeding flow rate of substrate solution through the RIB on conversion of the dextrin, it was shown that the maximal conversion was obtained at very low rates, not higher than 1–3 ml/min (Fig. 4). Under these conditions, the conversion of 1% and 10% corn dextrin were ca. 100% and 25-30%, respectively, that corresponded to the formation of glucose syrup and starch treacle, respectively.

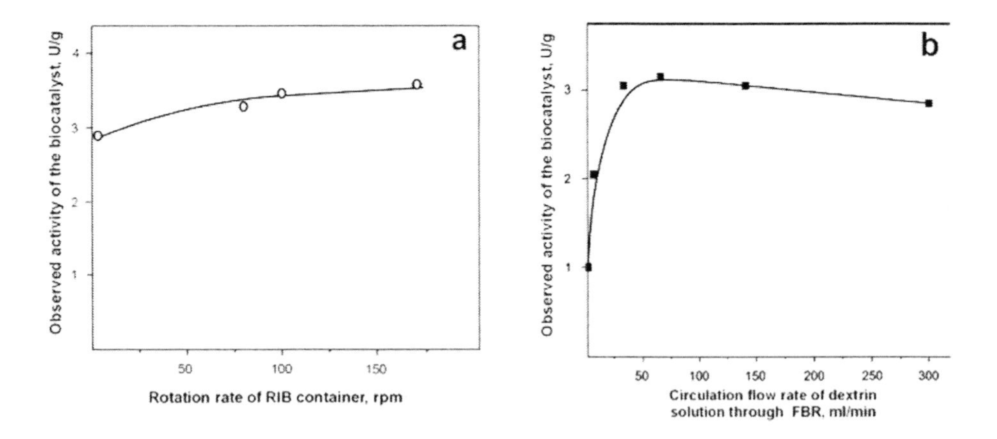

Fig. 3. Dependence between the observed rate of dextrin hydrolysis and the container rotation rate in RIB with a periodic mode (a) and circulation feeding flow rate of substrate solution through the biocatalyst bed in FBR (b). *Hydrolysis conditions* were following: 50°C, pH 4.6, and concentration of corn dextrin 1w/v%

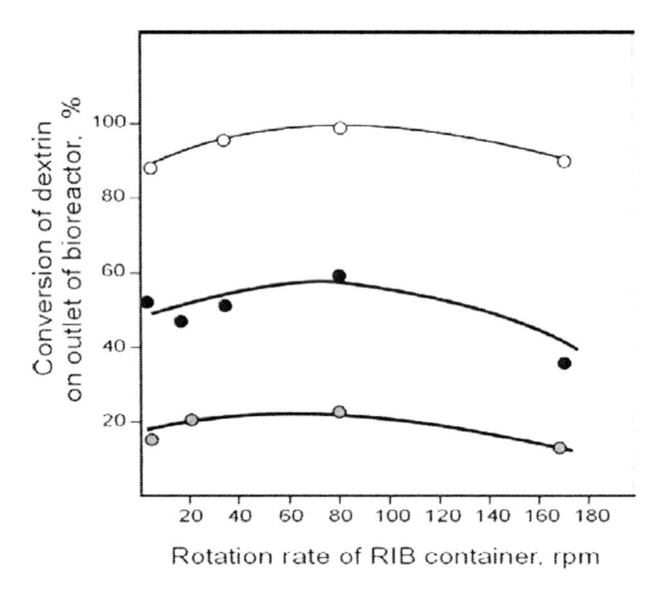

Fig. 4. Dependence between corn dextrin conversion and the feeding flow rates of substrate solution at 3 ml/min (o), 17 ml/min (•) and 70 ml/min (◉) in the RIB with a continuous mode. *Hydrolysis conditions* are following: 50°C, pH 4.6, and concentration of corn dextrin 1w/v%

In order to compare efficiency of the RIB and FBR, operating with the continuous mode, a parameter of "contact time", which was determined as a ratio of catalyst volume V (cm^3) to volumetric feeding rate υ (ml/min) was used. As follows from Fig. 5, at long contact time, the efficiency of the dextrin hydrolysis in the RIB was 1.2-2 times higher due to intensification of mass transfer and overcoming diffusion limitations.

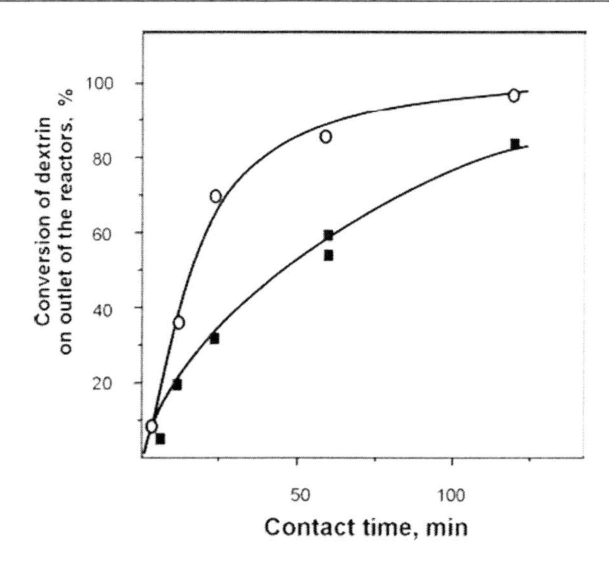

Fig. 5. Comparison of process efficiency of dextrin hydrolysis in RIB (o) and FBR (■). *Hydrolysis conditions* were following: $50^{O}C$, pH 4.6, concentration of corn or wheat dextrin 1w/v%, and rotation rate of container 170 rpm

The rotor-inertial bioreactor was also studied in the heterogeneous biocatalytical process of sucrose inversion. The biocatalyst for this process was prepared by adsorptive immobilization of baker yeast membranes on the carbon-containing foam-like ceramics. A comparison of the RIB and FBR showed that their efficiency was similar probably because of absence of diffusion limitations for the sucrose inversion process in study.

Vortex immersed reactor (VIR). *This reactor operated with a periodic mode only. As mentioned above, this regime was compared to the circulation mode of the fixed-bed reactor.*

As in the case of the rotor-inertial bioreactor, the VIR design was also aimed at preventing the formation of stagnant zones and significant intensifying mass transfer of a substrate to an enzyme immobilized on the porous support. In contrast to the RIB, the heterogeneous biocatalysts for VIR were prepared using granular (not foams) carbon-containing supports with a developed surface area (Table).

To prevent formation of stagnant zones, the profiled reactor body was designed to direct a substrate solution through biocatalyst bed. Since the profile of each body scuff plate followed the hyperbolic law, the annular cross-section area for a substrate solution stream was equal to $2\pi R \cdot h$ and remained constant for all time. It is well known that under these conditions of liquid flow, the stagnation or jet formation do not occur. As mentioned above, a substrate solution was sucked through a bottom hole upon the body rotation, and then moved through the biocatalyst bed toward side holes. During this movement, the liquid was obviously affected by hydrodynamic, centrifugal and inertial (Coriolis) forces, which resulted in a vortex flow of the substrate solution. As a result, intensification of mass transfer of the substrate to the porous biocatalyst was provided. The centrifugal forces were responsible for slight compression of the biocatalyst layer, narrowing of the channels for liquid flowing between granules, which result in the additional increase of mass transfer. Because of high mechanical strength of the inorganic supports for enzyme immobilization, the granules were not distorted during the reactor operation.

A lab-scale setup of the vortex-immersed reactor was studied in the heterogeneous biocatalytical process of starch dextrin hydrolysis. The glucoamylase activity and stability of the biocatalysts for this process were very high (Table). Thus the half-life time ($t_{1/2}$) exceeded 10 months upon long storage at 18-22°C (see Table) and seven hundred hours upon continuous running of the VIR at 50°C.

To elucidate optimal conditions for the VIR operation during the dextrin hydrolysis, the effect of the reactor body rotation rate on the hydrolysis was investigated. For all biocatalysts the observed biocatalytical activity remained constant at the rotation increased above 300 rpm (Fig. 6, a), and thus the external diffusion restrictions were overcome.

Studying the rate of dextrin hydrolysis *versus* size of the biocatalyst granules, it was shown that the process ran deeply under the internal diffusion limitation. Thus if Sibunit granules were grinded from 3 to 0.2 mm in size, the observed hydrolysis rate increased by an order of magnitude (Table, Fig. 6, a). Further a highly active biocatalyst prepared by adsorption of glucoamylase on the small granules of Sibuinit (ca.0.7 mm) was used in study.

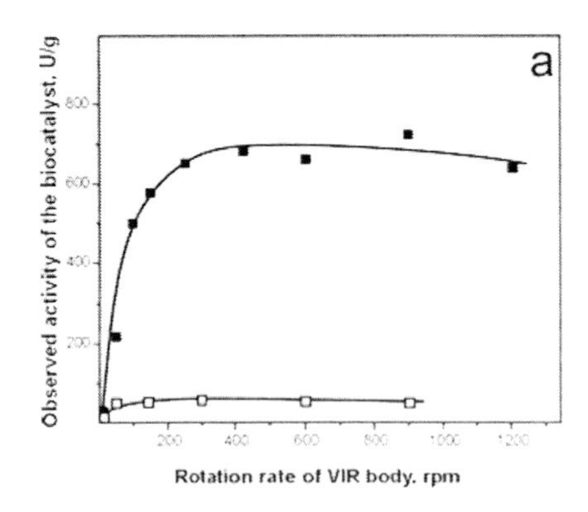

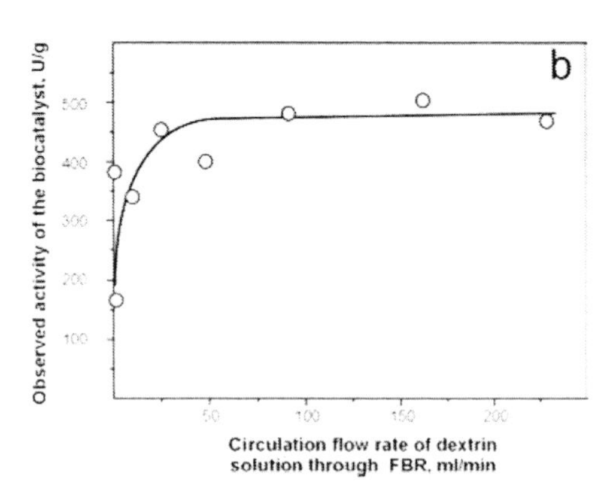

Fig. 6. Dependence berween the observed dextrin hydrolysis rate and rotation rate of the VIR body (a) filled by biocatalysts based on glucoamylase immobilized on Sibunit granules of 1,2-1 mm size (■) and 2-3 mm size (□) and circulation feeding flow rate through FBR (b). *Hydrolysis conditions* were following: 50°C, pH 4.6, and concentration of potato dextrin 10w/v%

Efficiency of heterogeneous processes of the dextrin hydrolysis was estimated from comparison of observed biocatalytical activities in the VIR and FBR when the external diffusion restrictions were overcome. As was shown, the condition were following: for VIR the body rotation rate (ω) was more than 450 rpm (Fig. 6, a), and for FBR circulation rate (υ) was more than 50 ml/min (Fig. 6, b). Note that the observed activity of the biocatalyst was higher in the VIR by 20-50%. In addition, it was found, that as height of the biocatalyst bed in FBR was increased, the hydrodynamic resistance of the biocatalyst bed increased due to its compression at high circulation rate.

As a result, rotor-inertial bioreactor (RIB) and vortex-immersed reactor (VIR) were designed in lab scale and studied for the heterogeneous hydrolysis of starch dextrin that was controlled by both external and internal diffusion of the substrate to the immobilized glucoamylase. It was shown experimentally that the external diffusion limitations were completely overcome at rotation rates of the RIB container and VIR body of $\geq$ 80 rpm and $\geq$450 rpm, respectively. The internal diffusion restrictions became minimal if the support granules were less than 1 mm. Thus, under these conditions the efficiency of the process of dextrin hydrolysis in vortex reactor was higher by 1.2-1.5 times than that in the fixed-bed reactor.

As a conclusion, the vortex reactors were developed for the heterogeneous biocatalytical processes. The vortex reactors outperform excellently the fixed-bed reactors if diffusion of the substrate was the limiting stage of the process. It occurred if a substrate of the enzyme was a high-molecular weight substance or if the heterogeneous biocatalyst possessed a very high enzyme activity. Also the design of vortex reactors provided full elimination of stagnant zones and jet formation in substrate solution.

REFERENCES

[1] G.Iosilevskii, H. Brenner, C.M.V. Moore, C.L. Cooney : *Phil.Trans.Roy.Soc.* London, 45(1675), 259 (1993).

[2] R.C. Giordano, R.L.C Giordan., D.M.F. Prazeres, C.L. Cooney*:* *Chem.Eng.Sci.*, 53 (20), 3653 (1998).

[3] R.L.C. Giordano, R.C. Giordano, D.M.F. Prazeres, C.L. Cooney : *Chem.Eng.Sci.*, 55 (18), 3611 (2000).

[4] R.L.C. Giordano, R.C. Giordano, C.L. Cooney: *Process Biochem.*, 36 (10), 1093 (2000).

[5] Morteza Sohrabi, Mehdi A. Marvast: *Ind.Eng.Chem.Res.*, 39 (6), 1903 (2000).

[6] G.A.Kovalenko, S.V. Sukhinin, A.V.Simakov, L.V.Perminova et al: *Biotechnology* (Moscow), 1, 83 (2004). (in Russian).

[7] Pat. 2245925 RU. Publ. 10.02.2005.

[8] G.A.Kovalenko, S.V. Sukhinin, L.V.Perminova *Biotechnology* (Moscow), 2005, in press. (in Russian)

[9] G.A. Kovalenko, O.V. Komova, A.V. Simakov et al: *J.Mol.Catal.A: Chemical.*, 182-183, 73 (2002).

[10] G.A. Kovalenko, V. A.Semikolenov, E.V.Kuznetsova et al: *Colloid J.*, 61 (6), 729 (1999).

[11] G.A. Kovalenko, E.V. Kuznetsova, Yu.I. Mogilnykh et al: *Carbon*, 39 (7), 1133 (2001).

In: Industrial Application of Biotechnology
Editors: I. A. Krylov and G. E. Zaikov, pp. 55-61
ISBN 1-60021-039-2

Chapter 7

DEVELOPMENT OF RESOURCE-SAVING TECHNOLOGY OF BOVINE PANCREATIC RIBONUCLEASE MANUFACTURE

M. M. Baurina, A. A. Krasnoshtanova, and I. A. Krylov*
Department of Biotechnology, D.I. Mendeleyev University of Chemical Technology of Russia, Moscow

ABSTRACT

The kinetics of extraction of ribonuclease (RNA-ase) from bovine or other cattle pancreatic gland has been studied. This stage of the technological process of RNA-ase manufacture has been investigated and the mathematical model of the process has been developed. The process of RNA-ase extraction has been shown to conform to the first-order equation. The optimum conditions of extraction have been determined: the temperature of 5°C, pH 2.5 and 15% bovine pancreatic cells (by dried weight) providing 60% output from its content in bovine pancreatic cells.

Key words: pancreatic ribonuclease, extraction, resource-saving technology, ENCADUM, yeast ribonucleic acid

INTRODUCTION

Pancreatic ribonuclease (RNA-ase) is one of the most investigated RNA-ases. RNA-ase isolated from bovine or other cattle pancreatic gland, purified according to the level of tentative pharmacopeia specification requirements is a medical product itself and it is also applied for producing another medical product – ENCADUM, pancreatic hydrolysate of yeast

* Department of Biotechnology, D.I. Mendeleyev University of Chemical Technology of Russia; 9, Miusskaya sq., Moscow, 125047Russia; fax: (095) 978-74-92; E-mail: krylov@muctr.edu.ru

ribonucleic acid, which is used in medicine for treatment for such diseases as progressive tapetochoroidal abiotrophy, neuromuscular, and other diseases [1, 2].

According to the well-known technology [3] RNA-ase is isolated from bovine pancreatic gland by extraction with 0.25 N solution of sulfuric acid at temperature of 0 - 5 °C. To purify the enzyme from accompanying proteins concentrated sulfate ammonium solution is used. Then the preparation is purified by recrystallization from ethyl alcohol, the yield of RNA-ase from pancreatic gland being 0.1% of the dried weight of the gland. Such a low yield is due to a large number of stages and low RNA-ase precipitability because of its low concentration in the solution.

The purpose of this investigation was to choose optimal conditions of RNA-ase extraction from native bovine or other cattle pancreatic gland cells providing the maximum of the enzyme yield on the base of quantitative regularities of the process considered.

Materials and Methods

The object of our investigation was native bovine or other cattle pancreatic gland cells, containing 20% of air-dry weight and 20% of protein compounds by dried weight. The initial activity of the cells was 16.6 units/mg (by dried weight). All reagents which provided realization of the researches were prepared with application of reactants of higher quality. The investigations of quantitative regularities of the protein extraction process were carried out in a glass apparatus of the volume up to 1000 ml provided with a water jacket to supply a cooling agent, an electrical mixer, a backflow condenser, and necks for sampling and thermometer setting.

On investigating the kinetics of protein extraction the definite amount of pancreatic gland biomass was brought in the sulfuric acid solution of known concentration and then mixed at constant temperature. The suspension was lodged in the apparatus that was cooled beforehand till the temperature required. The moment of the ending of cooling (the time of cooling did not exceed 5 minutes) was chosen as a moment of the beginning of an experiment, in the course of which the samples of the extract were taken for tests in certain intervals of time. The pancreatic gland cells were separated by centrifugation at 5700g for 10 minutes at 4°C.

The contents of two protein groups in the samples were determined by modified Lowry method, allowing to determine the concentrations of high molecular protein fraction (HMF), precipitated by 50% solution of trichloroacetic acid (TCA) and low molecular fraction (LMF) that is not precipitated by TCA.

Evaluation of the experimental data was carried out with the use of standard programs of integration of the differential equations and optimization of their parameters followed by comparison of the results of calculation with the experimental data according to Fisher's test, being used for check of adequacy of the developed mathematical model ($F_{opt} < F_{tab}$).

The investigation of quantitative regularities of the extraction was carried out according to three series of kinetic curves in each of which one of three factors that were able to influence on the enzyme extraction rate such as the temperature, pH of the solution, and the concentration of cells varied within the ranges: 5 – 20 °C, 0.075 – 0.25 mol/liter, 2.5 -10% by dried weight, respectively, was changed.

Results and Discussion

At the first stage of investigation the influence of the acid extractive solvent (sulfuric acid) concentration on the yield of proteins and RNA-ase from pancreatic gland has been studied.

In Fig. 1 typical curves of the accumulation of HMF (x_p) and LMF (x_{pp}) in the extract depending on the acid extractant concentration, the initial pancreatic gland cell concentration and the temperature are given in the plot of the extraction extent (x) vs. the extraction time (τ). To find out the kinetic relationships of RNA-ase and protein extraction the changes of specific HMF protein activity (unites/mg of the protein) vs. the time of the extraction have been determined. It was established that the specific protein activity is not changed at the time of extraction that allowed us to draw a conclusion that RNA-ase accumulates cymbately with the HMF of proteins. Therefore, the further investigations were to find out the optimum conditions of HMF protein extraction.

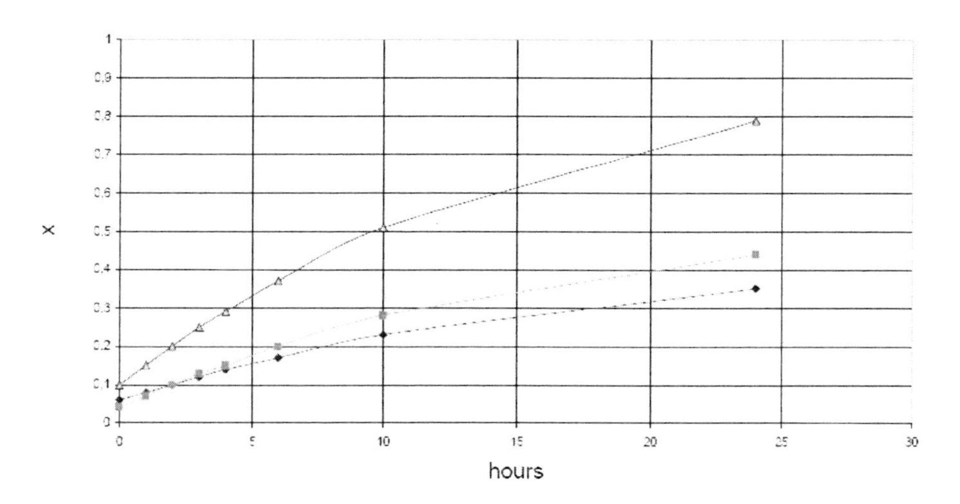

FIG. 1. Accumulation of the protein compounds in the extract: high molecular fraction (1), low molecular fraction (2), and total protein compounds (3) on the processing of bovine pancreatic gland cells with water solution of sulfuric acid. Conditions: the temperature of 5°C, the concentration of sulfuric acid in the medium is 0.1 mol/l (pH 2), the initial concentration of bovine pancreatic cells is 10% by the dried weight

The absence of dependence of extraction extent from the initial content of cattle pancreatic gland cells by dried weight has allowed us to assume the 1-st order on the solid phase concentration. At previous studying of the acid protein extraction from the yeast and bacteria [4] the pathway of serial-parallel transformations of intracellular proteins and their accumulation in the extractant was suggested. With reference to our case there were no bases to refuse such a pathway. However, taking into account the absence of S-shaped behavior of the kinetic curves it is possible to disregard a stage of intracellular intermediate formation that has allowed us to use the following pathway of transformations:

$$\begin{array}{l} k_p \; k_{pp,1} \\ P_N \rightarrow P_1 \rightarrow PP_1 \\ \downarrow k_{pp,0} \\ PP_{0,} \end{array} \qquad (1),$$

where: P_N – intercellular proteins; P_1 – high-molecular proteins, formed in the solution; PP_0 and PP_1 – low-molecular products of protein hydrolysis; k_p, $k_{pp,0}$, $k_{pp,1}$ – the effective constants of the rate of single stages, correspondingly.

At the fist stage of investigation the processing of the experimental data on initial rates was carried out. As a result, the linear dependence for which the coefficient of correlation amounts as much as $r_k = 0.95$ (Figure 2) in the plot of ln P_{N0} against τ that has allowed with the slope ratio to determine the value of the sum of the effective constants (k_p and $k_{pp,0}$) has been obtained.

To determine separately the constants k_p and $k_{pp,0}$ the relationship obtained in the plot of Δx_P vs. $\Delta x_{PP,0}$ (Fig. 3) was used where Δx - difference in current and reference values of appropriate protein fraction extraction degrees on initial sites of kinetic curves. In the kinetic scheme (1) for parallel transformation of the first order the slope of the initial linear site of the line against the x-axis has been $k_p/\ k_{pp,0}$. The correlation of the experimental data processing results (Fig. 2 and 3) allowed us to estimate numerical values of constants k_p and $k_{pp,0}$. Then the value of the constant k_{pp1} has been found by fitting at integration of the differential equation system of the pathways of transformations (1).

The following stage was to study the influence of the temperature and the acid concentration on the process of protein extraction from pancreatic gland cells. The series of kinetic curves for various temperatures and concentrations of sulfuric acid were processed with the above described approach.

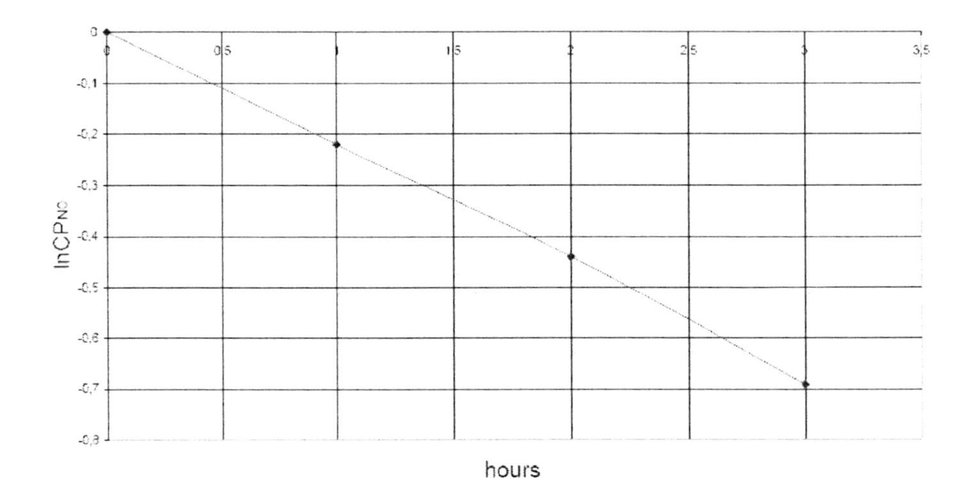

FIG. 2. Relation of logarithm of initial protein concentration vs. the time; Conditions: the temperature of 5°C, the concentration of sulfuric acid in the medium is 0.1 mol/l (pH 2), the initial concentration of bovine pancreatic cells is 10% by the dried weight

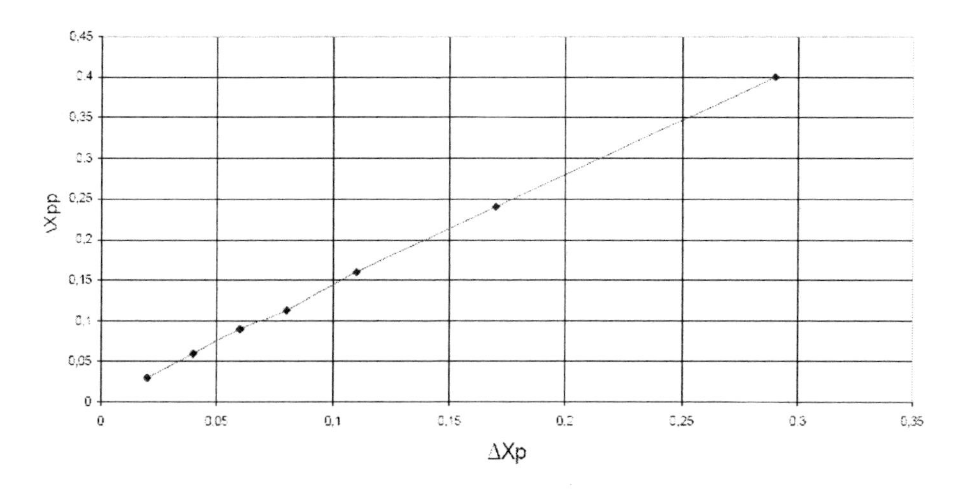

Fig. 3. Relation of differences in current and reference values of appropriate protein fraction extraction degrees on initial sites of kinetic curves; Conditions: the temperature of 5°C, the concentration of sulfuric acid in the medium is 0.25 mol/l (pH 2), the initial concentration of bovine pancreatic cells is 10% by the dried weight

At studying the influence of the temperature on the rate of protein extraction Arrhenius equation was used:

$k_{ef} = A_0 \cdot e^{-E/RT}$	(2)

and at studying the influence of the acid - the equation of specific acid-base catalysis:

$k_{ef} = k^*/(1+K_b^a \cdot h_0)$	(3)

where h_0 is acidity of the sulfuric acid solutions, k* is the effective constant of rate that is not depending on acidity of the medium, K_b^a is the equilibrium constant responsible for forming appropriate intermediates at protonation of peptide bonds.

The obtained values of effective constants k_{ef} corresponding to above mentioned series of kinetic curves were used for calculation of numerical values of a preexponential factor (A_0) and activation energy (E) in equation (2) and parameters of equation (3), investigating the relationships in the plots ln k_{ef} vs. T^{-1} and k_{ef}^{-1} vs. h_0^{-1} (Fig. 4 and 5). The results of these researches have shown that in each case the linear dependences that confirm validity of the above chosen equations are observed with the value of correlation coefficients $r_k \geq 0.95$. It has allowed us to calculate the values of parameters for equations (2) and (3) which are given below in Table 1.

Thus, the scheme of the transformations (1) and equations (2) and (3) are determined on the whole as a mathematical model of the process of protein extraction from pancreatic cells which describes experimental data adequately. Its adequacy proves to be true with $F_{exp} = 3.74 < F_{tab} = 5.35$, where F_{exp} and F_{tab} are experimental and tabulated values of Fisher's test for significance level 0.05.

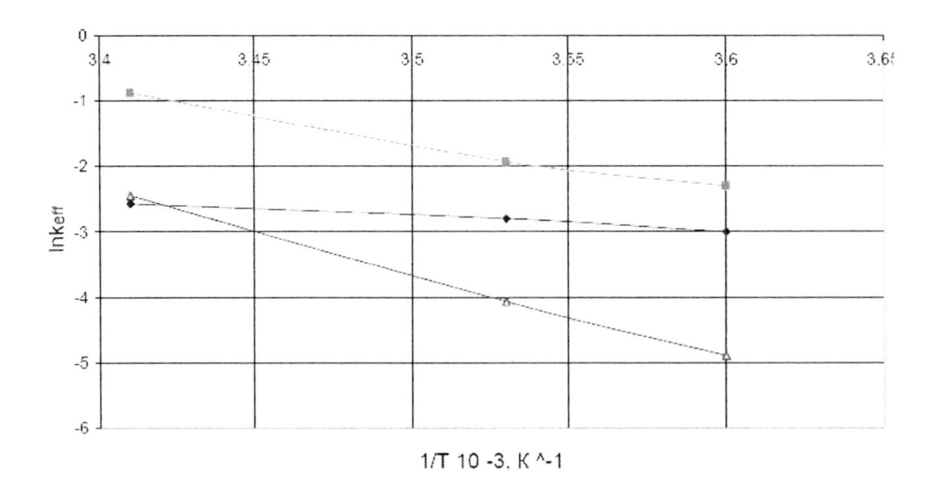

FIG. 4. Relation of logarithms of effective constants of each stage of transformation in the scheme of the acidic extraction of protein compounds from bovine pancreatic cells vs. reciprocal value of absolute temperature: 1, 2, 3 – the behavior of the constants k_p, $k_{pp,0}$, and k_{pp1} in the process of extraction of protein compounds from bovine pancreatic cells; Conditions: the concentration of sulfuric acid in the medium is 0.1mol/l (pH 2)

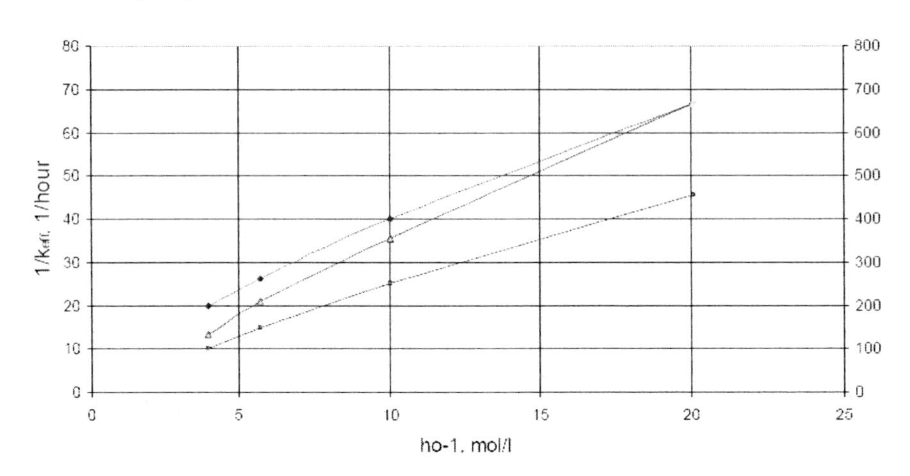

FIG. 5. Relation of logarithms of effective constants of each stage of transformation in the scheme of the acidic extraction of protein compounds from bovine pancreatic cells vs.the acidity of the medium h0 (double-reciprocal plot): 1, 2, and 3 – the behavior of the constants k_p, $k_{pp,0}$, and k_{pp1} in the process of extraction of protein compounds from bovine pancreatic cells; Conditions: the temperature of 5°C

Table 1. The parameters in equations (2) and (3) for the process of RNA-ase extraction from cattle pancreatic gland cells by individual stages of scheme (1)

The stage of transformation, determined by the effective constant in scheme 1	K_b^a, l/mol	$k^{*}\cdot 10^{-3}$, sec^{-1}	A_0, sec^{-1}	E, кJ/mol
k_p	3.34±0.27	1.7±0.13	0.14±0.009	12±0.91
$k_{pp,0}$	1.49±0.12	5.2±0.37	163±13.2	27±1.94
$k_{pp,1}$	1.52±0.11	0.33±0.028	63000±5680	45±3.87

*) – the value of the constant independent of the pH medium value at the temperature of 5°C

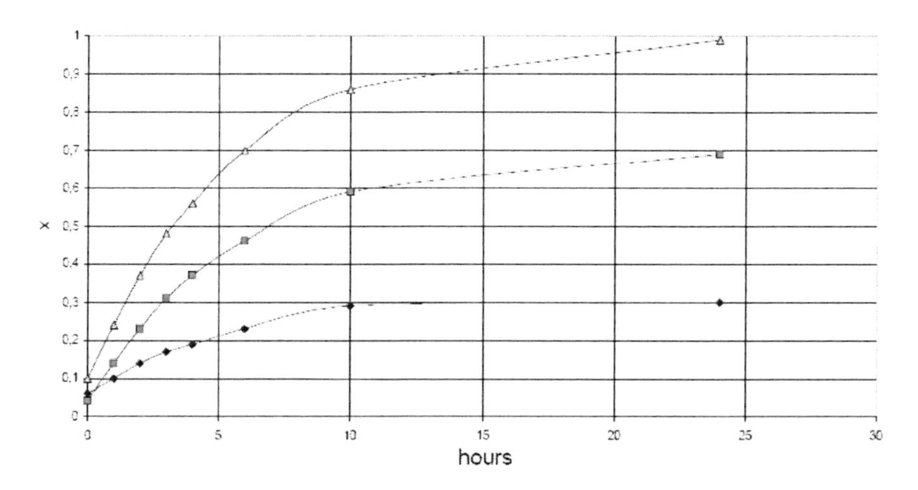

FIG. 6. Extraction of the protein compounds from bovine pancreatic gland cells in optimum conditions: high molecular fraction (1), low molecular fraction (2), and total protein compounds (3) on the processing of bovine pancreatic gland cells with water solution of sulfuric acid; Conditions: the temperature of 8°C, the concentration of sulfuric acid in the medium is 0.25mol/l (pH 2)

Using the mathematical model given, the optimum conditions of RNA-ase extraction from pancreatic cells were designed. As RNA-ase is extracted from the pancreatic cells cymbately to HMF of proteins the optimization was carried out by determination of maximum yield of LMF. The results of calculations further confirmed experimentally have shown that the technologically acceptable maximum of yield of RNA-ase (60% from the total mass of the enzyme in the pancreatic cells) with 1364 units/mg of specific activity is reached provided the processing of cattle pancreatic gland cells (containing 20% by dried weight) is carried out at 0.25 mol/l sulfuric acid concentration during 10 hours at the temperature of 5°C with the following double washing on the filter with acid solution (pH 2.0) taken in amount of 30% from sludge volume.

REFERENCES

[1] B.B. Fux, M.M. Shabanova, S.V. Fedorov, et al.: Method of preparing a mixture of ribonucleotides. *Patent Number* 5,064,758 (USA), C12N, 1991.

[2] V.M. Zemskov et al. (Ed.): *Low Molecular RNA – Production, Hydrolysis and Uses in Medicine.* Zinatne, Riga, 1985. 191p. (in Russian).

[3] M. Mcdonald: *Ribonucleases. Methods in enzymology.* Academic Press. New York, vol2, p.427-436 (1955).

[4] I.A. Krylov, A.A. Krasnoshtanova, and M.N. Manakov: Acid Hydrolysis of Protein Substances of Biomass of Industrial Microbial Producers. *Biotechnology*, №3, Moscow (1996).

In: Industrial Application of Biotechnology
Editors: I. A. Krylov and G. E. Zaikov, pp. 63-72
ISBN 1-60021-039-2

Chapter 8

INVESTIGATION OF QUANTITATIVE REGULARITIES OF PROCESS OF NADH AND RNA EXTRACTION FROM BAKERY YEAST

R. A. Parizheva, A. A. Krasnoshtanova and I. A. Krylov*
D.I. Mendeleyev University of Chemical Technology of Russia
Moscow, Russia

ABSTRACT

Conditions of extraction of RNA and reduced forms of NADH from yeast *Sacharomyces cerevisiae* biomass were determined. It was established, that the optimum conditions of RNA and reduced nicotinamide coenzyme extraction from bakery yeast biomass are the temperature 80 °C; pH 8.0, the time of extraction 20 minutes, and sodium sulphite concentration 0,5 mol/l, the RNA output being 70 %, and the reduced nicotinamide coenzymes output being 91.2 %.

Key words: nicotinamide coenzymes, RNA, extraction, bakery yeast.

INTRODUCTION

Nicotinamide coenzymes draw attention of physiologists, biochemists and clinical physicians. It is established that these nucleotides being components of many coenzyme systems play an important role in vital processes. The basic role of nicotinamide coenzymes consists in their participation in electron and hydrogen transportation from oxidizing substrata to oxygen in a cellular respiratory chain. Nicotinamide adenine dinucleotide is necessary for anaerobic way of formation of high energy phosphoric compounds. It can account for the presence of nicotinamide coenzymes practically in all types of live cells. Nicotinamide

---

* D.I. Mendeleyev University of Chemical Technology of Russia. 9, Miusskaya sq., Moscow, 125047, Russia. E-mail: krylov@muctr.edu.ru

adenine dinucleotide is used in medicine as immunomodulator. Nicotinamide adenine dinucleotide provides a neurotransmitter function – a transfer of nervous impulses between the brain and other parts of the body.

Nucleic acids of low and average molecular weights draw attention of biotechnologists as perspective sources of raw material for manufacture of nucleosides and 5'-mononucleotides, making a basis in synthesis of medical products used in medical treatment for a wide range of diseases. It is microbic RNA and DNA advantage over other sources that their hydrolysis results in formation of four basic nucleosides or 5'-mononucleotides which are separated by ionic exchange to yield the crystal preparations used in synthesis of medicinal forms.

In industry nicotinamide coenzymes are produced by microbiological synthesis by *Brevibacterium ammoniagenus* [1].

As a rule, to increase an output of nicotinamide coenzymes the predecessors are added such as adenine, nicotinic acid amine in a nutrient medium at cultivation of microorganisms. The output of nicotinamide coenzymes does not exceed 0.1 % in terms of dried weight of bacteria that makes such manufacture unprofitable. There are also chemical ways of synthesis of nicotinamide coenzymes, but however, all of them assume the use of expensive reagents and multiphase purification of end-products. Therefore they may be used only for preparative purposes [2].

Complex processing of bakery yeast biomass with extraction of ribonucleic acid (RNA) is more perspective way of nicotinamide coenzyme manufacture [3].

It is expedient to use water-alkaline or water-salt solutions with pH 8 - 9 for RNA extraction from yeast. Reduced forms of nicotinamide coenzymes having the greatest practical value have got the maximal stability in these conditions. Therefore it is represented actual to develop a technology of complex processing of bakery yeast biomass with parallel isolation of yeast RNA preparations and reduced nicotinamide coenzymes (NADH and NADPH).

The key stage of the technological process is nicotinamide coenzyme and RNA extraction one. Therefore the purpose of the research work is to study quantitative regularities of nicotinamide coenzyme and RNA extraction from bakery yeast biomass.

METHODS AND MATERIALS

The bakery yeast biomass containing 25 % of dried substances was used. The biomass contains 6 % of nucleonic acids and 0.264 % of reduced nicotinamide coenzymes by dried weight.

Research of quantitative regularities of nicotinamide coenzyme and RNA extraction from bakery yeast biomass was carried out in the glass reactor of the volume of up to 1000 ml provided with a mixer and several necks for sampling and also supplied with the electric drive, the thermometer, a water jacket, and electrodes of a pH-meter.

At studying the kinetics of extraction the following order of realization of the experiment was used: yeast suspension prepared beforehand was brought in a reactor warmed up to the temperature required, and the required value of pH was established. The moment of the ending of heating suspension (the time of heating did not exceed 5 minutes) was accepted as the beginning of extraction process. Through the certain time intervals the samples of the extract were taken for tests. The samples were cooled up to room temperature, and then the biomass was separated by centrifugation. In the cell-free extracts the contents of nucleonic components were determined by A. Spirin method, the contents of reduced nicotinamide coenzymes being defined by the absorption at the length of wave as much as 340 nm, and oxidized ones being determined with technique [4].

NADH, NAD alcohol dehydrogenase, mandelic acid, and cysteine used in this research work were purchased from ICN. All reagents which provided realization of analyses were prepared with application of reactants of higher quality.

Evaluation of the experimental data was carried out with the use of standard programs of integration of the differential equations and optimization of their parameters followed by comparison of the results of calculation with the experimental data according to Fisher's test, being used for check of adequacy of the developed mathematical model ($F_{opt} < F_{tab}$).

RESULTS AND DISCUSSION

At the given investigation phase the problem was to choose the optimum conditions of extraction, providing the maximal output of reduced nicotinamide coenzymes and oxidized ones (NAD and NADP) without division.

Research of quantitative regularities of nicotinamide coenzyme extraction from bakery yeast biomass was started with obtaining series of the kinetic curves for the various initial contents of cells in suspension, the temperature and pH of the medium. The specified parameters were varied in intervals: 60 - 90°C; 2.5 - 10 % of dried substances and pH 7.0 - 10.0 accordingly.

Typical curves of nicotinamide coenzyme extraction from bakery yeast biomass received at the various initial contents of yeast are given in Figure 1. Within the limits of accuracy of measurements the coenzyme share does not depend on the initial cell contents in suspension. The given fact specifies that the process researched at constant initial concentration of alkali in the medium may be described by the kinetic equation of the first order on processed yeast mass. Besides accumulation of the oxidized form of coenzyme in a solution, and also absence of points of inflection on initial sites of kinetic curves have allowed assuming the following pathways of series-parallel transformations for processing experimental data:

$$(NAD(P)H)_S \xrightarrow{k_{21}} (NAD(P)H)_L \xrightarrow{k_{22}} (NAD(P))_L \qquad (1)$$

$$(NAD(P)H)_S \xrightarrow{k_{20}} (NAD(P))_L$$

where: $(NAD(P)H)_S$ and $(NAD(P)H)_L$ - concentration of reduced nicotinamide coenzymes inside a cell and in volume of a liquid phase, NAD – concentration of NAD in volume of a liquid phase, and k_{20}, k_{21}, k_{22} - effective constants of the investigated process, dependent on the temperature and concentration of alkali, accordingly.

The pathway of reaction transformations (1) offers the first order of the reaction on soluble (NAD(P)H), and each of stages submits to acid-basic catalyst regularities. Hence, the following sequence of processing experimental data was chosen.

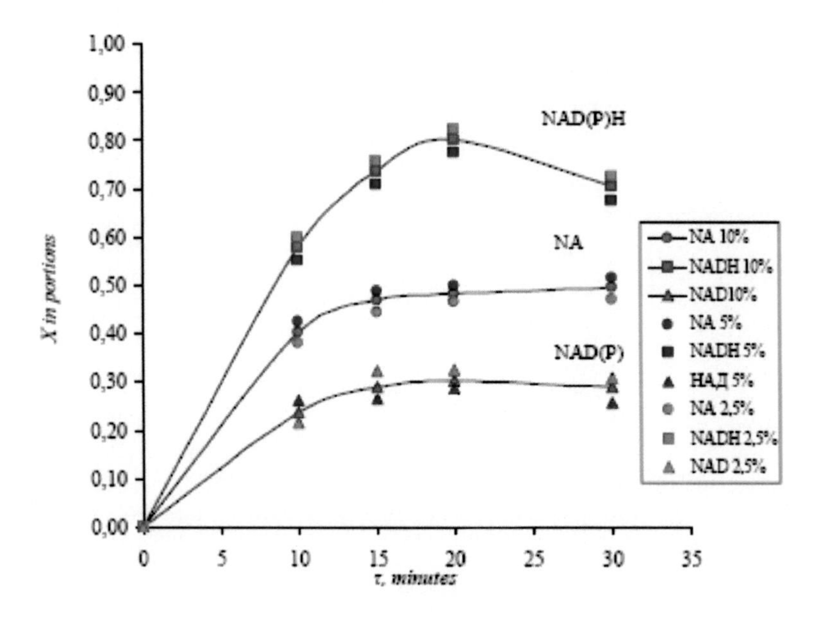

Figure 1. Typical curves of RNA extraction, the reduced and oxidized forms of nicotinamide coenzymes from biomass bakery yeast. Conditions: the temperature 80°C; pH 8.0

At the first stage of studying researched process kinetics the processing on initial speeds of a series of the kinetic curves received at constant temperature and pH of medium was carried out.

Apparently from Figure 2, the linear function is observed in coordinates ln(1 - x) – τ (min) with factor of correlation $r_к = 0.95$ where x – a degree of reduced nicotinamide coenzyme extraction. The tangent of the inclination corner of a straight line to an abscissa axis corresponds to the sum of constants k_{20} and k_{21}.

It is possible to estimate numerical values of each of constants k_{20} and k_{21}, using a linear function of NAD (P), NAD (P) H and NAD (P) extraction degree that is typical for initial sites of all set of kinetic curves (Figure 3). Presence of such site specifies to the course of parallel chemical processes of the identical order that is quantitatively characterized by constants k_{20} and k_{21}. Thus, the tangent of the linear part curve inclination corner to the abscissa axis is equal to the relation k_{20}/k_{21} (Figure 3). As the type of the function submitted in Figure 3 is typical for all kinetic curves of nicotinamide coenzyme extraction from yeast it allows to estimate the values of constants k_{20} and k_{21} for each kinetic curve. On increasing the time of extraction the curve in Figure 3 aspires to saturation that specifies at the appreciable contribution of formation of oxidized nicotinamide coenzyme forms on a consecutive way which is characterized by constant k_{22} in the pathways of transformations (1).

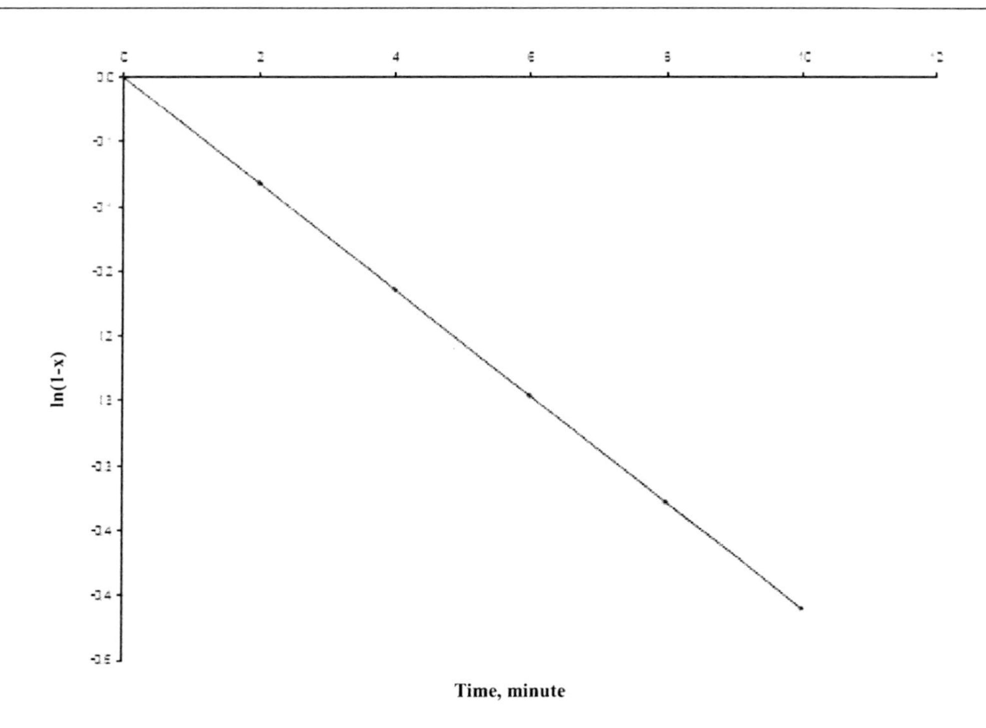

Figure 2. Processing of experimental data on extraction of nicotinamide coenzymes from biomass bakery yeast on initial rates. Conditions: the temperature 80°C; pH 8.0, the concentration of the yeast biomass -10 % by dried weight

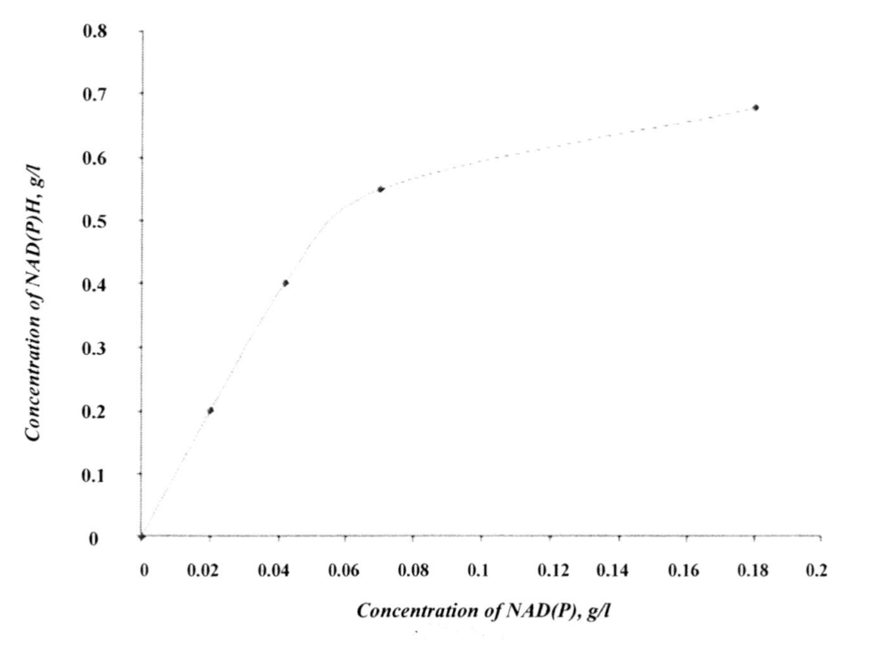

Figure 3. Dependence of concentration of the oxidized forms of nicotinamide coenzymes and concentration of reduced ones Conditions: the temperature 80°C, pH 8.0, the concentration of yeast biomass 10% by dried weight

The value of constant k_{22} is established by numerical selection of values during integration of system of the differential equations adequate to the pathway of transformations (1).

The data processing has allowed estimating numerical values of all constants in the considered pathways of transformation (1) that further has allowed solving the task for their optimization as parameters of the differential equation system. At first the values of constants k_{20} and k_{21} were optimized on the initial sites of appropriate kinetic curves, and then the values of constant k_{22} were specified on the average and final sites of curves.

The following stage of processing experimental data provided finding the parameters of equations of the relationship between effective values of constants k_{20}, k_{21} and k_{22} with the temperature of extraction and alkaline extractant concentration.

The influence of the temperature on the nicotinamide coenzyme extraction rate was established by processing a series of the kinetic curves received at the constant initial contents of yeast and sodium hydroxide in suspension.

Taking into account a rather narrow interval of the temperature change the Arrenius equation was offered to use. As the results of calculations have shown, in coordinates ln k_{effect} - T^{-1}, where T – the absolute temperature, it can be regarded as a linear one. The correlation coefficient has seemed to be high enough (not less 0.95). It has allowed calculating the appropriate values of activation energy and a preexponential factor for each stage of the pathways of transformation (1). The results of calculations are given in Table 1.

Similarly, the influence of various concentrations of sodium hydroxide on the extraction rate was investigated. Here, the specific acid-base catalysis equation which assumes equilibrium formation of a complex between nicotinamide coenzymes and hydroxide ions with constant K_B (a fast stage) and hydrolysis of the later with constant k_0 (a slow stage) was used for finding the relationship between the values of the effective constants corresponding to a series of kinetic curves received at the constant temperature and constant yeast suspension concentration, but with various concentrations of the alkaline agent. Taking into account a low initial sodium hydroxide concentration in the medium, the activity of a water solution of alkali (a_{OH}) was accepted instead of a parameter of basicity. Then the equation of relationship between effective constants and sodium hydroxide contents in the solution is:

$$k_{eff} = k_0 \cdot a_{OH} / (K_B + a_{OH}) \qquad (2)$$

It was possible to find numerical values of parameters k_0 and K_B in equation (2) as a result of processing the appropriate experimental data in coordinates $k_{эф}^{-1}$, a_{OH}^{-1}. Apparently from Figure 4, in above mentioned coordinates the functions observed can be considered to be linear ones ($r_к > 0.94$).

The values of parameters K_B and k_0 and the appropriate deviations are submitted in the Table given below.

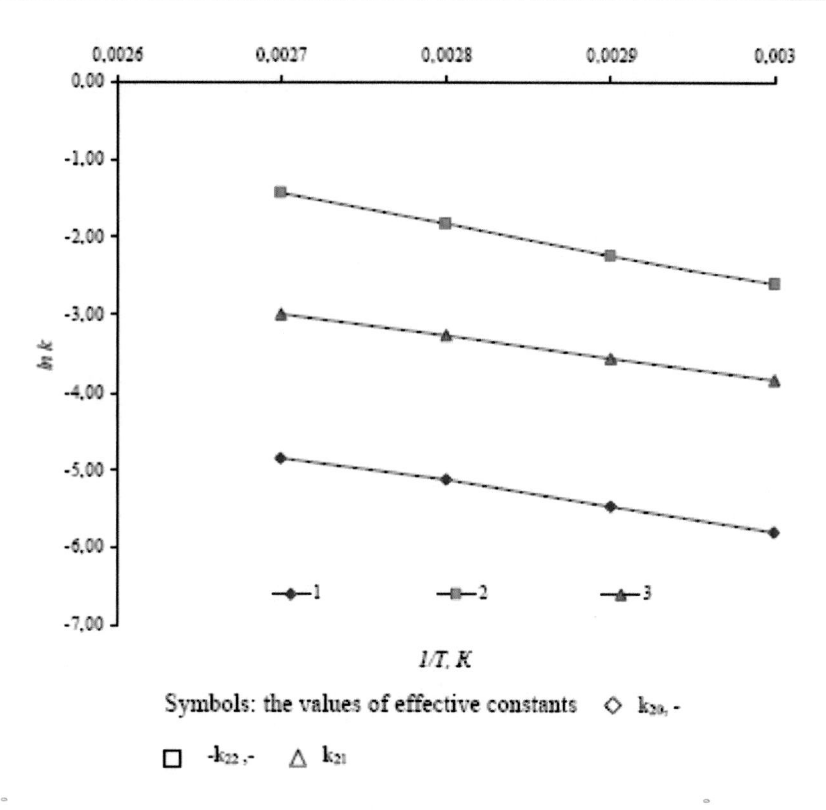

Figure 4. Dependence of effective constants of rates of extraction of reduced and oxidized forms of nicotinamide coenzymes from activity hydroxide ions in vs. coordinates. Conditions: the temperature 80°C, the concentration of a yeast biomass 10 % by dried weight

Table 1. Values of effective constants of process of extraction of nicotinamide coenzymes from biomass of bakery yeast

Pathway	E, кJ/mol	A_0, min $^{-1}$	k*, min^{-1}	K_B, l/mol
k_{20}	15,0 ± 0.9	0,012 ± 0.0007	0.001± 0.0001	0,0012 ± 0.00010
k_{21}	6,0 ± 0.4	0,042 ± 0.0025	0.027 ±0.0021	0,0017 ± 0.00010
k_{22}	10,0 ± 0.6	0,061 ± 0.0037	0.175 ± 0.0120	0,0002 ± 0.00001

Thus, the pathways of transformations (1), Arrenius equation, equation (2) and numerical values of their parameters define a mathematical model of nicotinamide coenzyme extraction which describes experimental data adequately. To select the optimum conditions of reduced nicotinamide coenzyme extraction the developed mathematical model was used. As the task of the given work is to develop a technology of complex processing baker yeast biomass with parallel RNA extraction a mathematical model of RNA extraction was used besides the developed mathematical model of nicotinamide coenzyme extraction [5].

The results of calculations confirmed experimentally have shown that the maximal output of reduced forms of nicotinamide coenzymes is observed at the temperature 80 – 90 °C; pH 8,0 - 9,0; the time of extraction 20 minutes is 78 - 83 %, however RNA extraction degree does not exceed 36 - 50 % in the same conditions.

The results of the subsequent calculations and experiments have shown that RNA extraction degree can be increased up to 90 % by increasing the time of extraction up to 60 minutes. However, the rate of nicotinamide coenzyme destruction also increases in such conditions; it results in decreasing their output till 25 - 30 %.

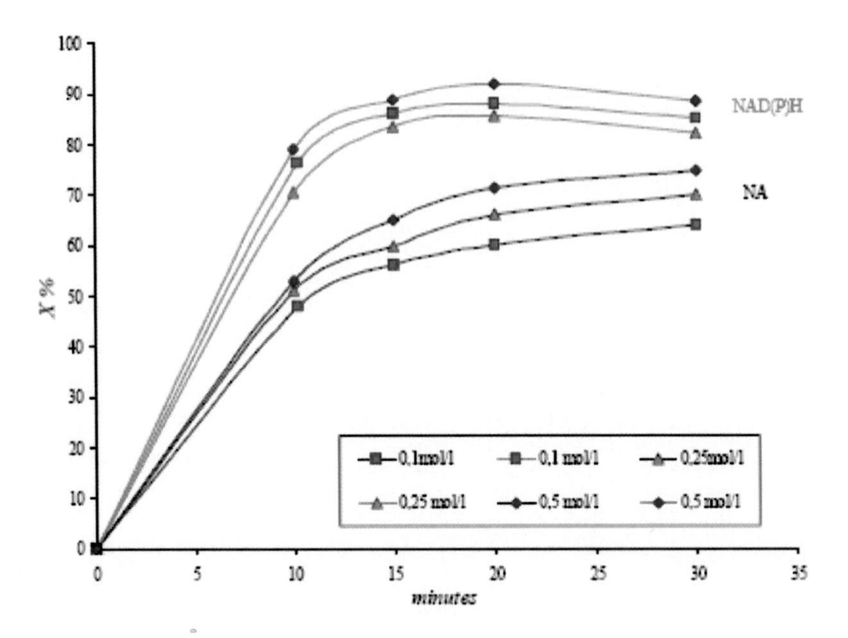

Figure 5. Dependence of a degree of RNA extraction (X) and a degree of extraction of the reduced forms of nicotinamide coenzymes from concentration of sulphite of sodium. Conditions: the temperature 80°C; concentration of yeast biomass -10 % by dried weight

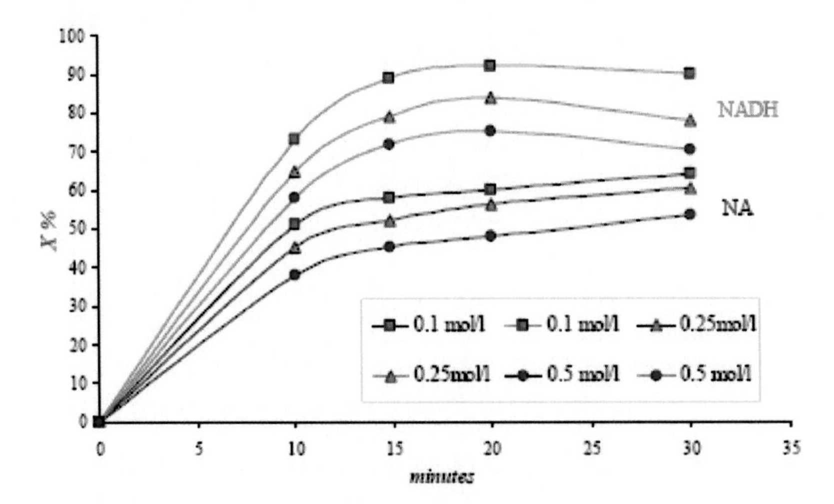

Figure 6. Dependence of a degree of RNA extraction (X) and a degree of extraction of the reduced forms of nicotinamide coenzymes from concentration of cysteine. Conditions: the temperature 80°C; the concentration of yeast biomass -10 % by dried weight.

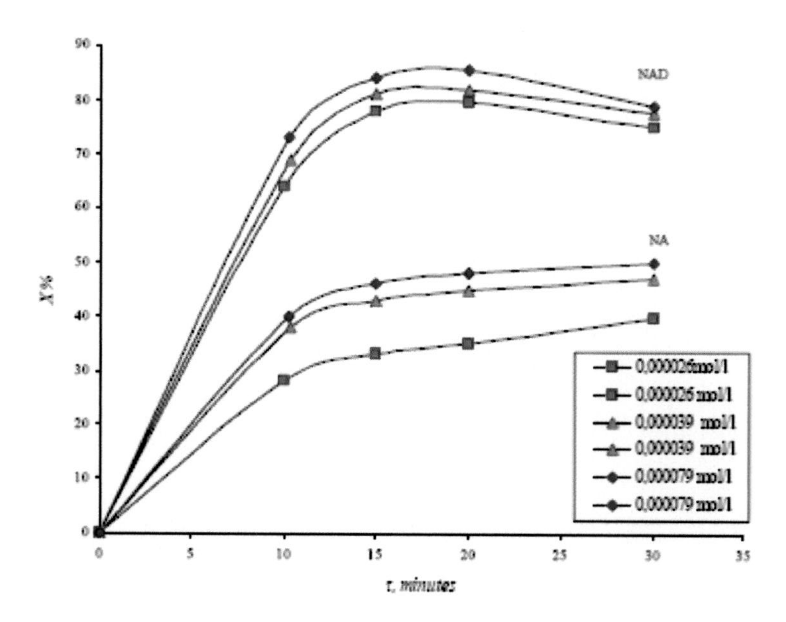

Figure 7. Dependence of a degree of RNA extraction (X) and a degree of extraction of the reduced forms of nicotinamide coenzymes from concentration of mandelic acid. Conditions: the temperature 80°C; concentration of yeast biomass -10 % by dried weight

To increase RNA output and to lower a nicotinamide coenzyme destruction degree at the same time, the influence of antioxidizers and nuclease inhibitors was investigated on the output of target products. Cysteine and sodium sulphite were used as antioxidizers, and mandelic acid was used as a nuclease inhibitor.

The results of experiments are submitted in Figures 5 - 7. The RNA output is stated to be 65 - 70 % in concentration of sodium sulphite and cysteine 0,5 mol/l and 0,1 mol/l in a reactionary mix respectively and pH 8.0, the temperature 80 °C and the time of extraction 20 minutes, reduced nicotinamide coenzymes being 90 - 92 %.

As the effect of the above-stated substances is identical, it is expedient to use sodium sulphite.

Thus, it was established, that the optimum conditions of RNA and reduced nicotinamide coenzyme extraction from baker yeast biomass are the temperature 80°C; pH 8.0; the time of extraction 20 minutes, and sodium sulphite concentration 0,5 mol/l, the RNA output being 70 %, and the reduced nicotinamide coenzyme output being 90 - 92 %.

REFERENCES

[1] *Industrial microbiology*. The main editor N.S.Egorov, Moscow, 1989 (in Russia).

[2] Synthesis of Nicotinamide Adenine Dinucleotide (NAD) from Adenosine Monophosphate (AMP): *J. of the American Chem. Society* 1980 - v. 102, 7805-7806 p. (in Russia).

[3] *Practical work on biochemistry*: The main editors S.E.Severin, G.A.Solov'evoj. - Moscow, 1989 (in Russia).

[4] N.V. Volkova.: *Development of technology microbic RNA and DNA from methane utilizable bacteria METHILOCOCCUS CAPSULATUS*. The author's abstract of the dissertation, Moscow, 1996 (in Russia).

In: Industrial Application of Biotechnology
Editors: I. A. Krylov and G. E. Zaikov, pp. 73-83
ISBN 1-60021-039-2

Chapter 9

DIAGNOSTICS OF AEROBIC PROCESSES OF BIOSYNTHESIS IN THE REAL TIME MODE

S. G. Mukhachev[*]
The Kazan State Technological University, Kazan

ABSTRACT

The system of operative analysis of aerobic growth and biosynthesis in phase space of metabolic speeds is developed. Energy-releasing and energy-consuming metabolic blocks are represented in the equations of a constructional exchange with weighting coefficients proportional to generation and consumption of energy. Since stoichiometric equations are linear, qualitatively different areas in phase space of metabolic speeds are separated by linear boundaries. On this basis the graphically clear form of displaying the information on process of an aerobic biosynthesis on the display of the operator-technologist workplace is built.

Keywords: an aerobic biosynthesis, a metabolism, a material balance, a gas balance, a phase space, a metabolic rate.

Methodical basis of operative diagnostics of an aerobic biosynthesis is considered by the example of growth of ethanol oxidizing yeast Candida lambica. The given process is chosen as the model which is characterized by replacement of various stoichiometrical variants of a metabolism, allowing to give a demonstration of characteristics of a method.

For the description of biosynthesis processes the following basic assumptions have been accepted:

(1) The metabolic system of a cell is in steady-state conditions, i.e. the magnitude of pool of energy rich compounds containing in single amount of a biomass is constant.

[*] The Kazan state technological university, 420015, Kazan, K.Marks street, 68. E-mail: biotex_kgtu@rambler.ru; Fax: 8432-367542

(2) Allocation of energy flows in a living cell can be appreciated according to standard S.J. Pirt's [1] postulate about additivity of consumption for maintenance of vital functions and growth of a cell.
(3) In consequence of the fact that cultivation of microorganisms is carried out practically in constant physical and chemical conditions, production of energy in a cell is possible to be considered dependent only from the intrinsic biochemical mechanism of transformation of a substrate molecule; influence on the specified process of small alteration of external conditions can be neglected.
(4) Energy consumptions on processes of biopolymers synthesis in a cell can be determined on the basis of the Guess's law, as equal to heat of biomass formation from carbon dioxide, ammonia and water: depth of rearrangement of a molecula of any substrate in a circuit of many hundreds transformations in the biochemical machine of a cell is so essential, that the process of a biosynthesis as a whole is equivalent to synthesis from "products of combustion" of a biomass [2].
(5) The liquid phase in the laboratory bioreactor is considered as homogeneous medium; distinctions between inside- and extracellular concentrations of substances are neglected.

The equations of balance are special cases of stoichiometrical invariants; that is why construction of invariant equations with elimination of separate components is possible. It allows to receive the supplementary constraint equations between amounts of the consumed substrates and synthesized products and to use them at operative calculation of cultivation processes balances.

The material balance of ethanol oxidizing yeast growth process can be represented by collection of the equations of formation and consumption of energy rich compounds (recalculated on ATP).

Essential simplification of the notation can be obtained under reduction (recalculation) formulas of all carbonic substances (compounds) to one atom of carbon.

The energy metabolism equation we shall represent in the form of:

$$CH_qO_r + \alpha_{O1} \cdot O_2 + E_S \cdot ADP \rightarrow \alpha_{C1} \cdot CO_2 + \alpha_{H1} \cdot H_2O + E_S \cdot ATP$$

Where: E_S – an energy rich compounds output from single amount of substrate under its full oxidation in recalculation on ATP.

Let's express oxygen consumption through a degree of substrate disoxidation γ_S. For this purpose we shall write down the equations of elements balance on carbon, hydrogen and oxygen: $1 = \alpha_{C1}$; $q = 2 \cdot \alpha_{H1}$; $r + 2 \cdot \alpha_{O1} = 2 \cdot \alpha_{C1} + \alpha_{H1}$.

Whence we shall express stoichiometrical coefficients: $\alpha_{C1} = 1$; $\alpha_{H1} = q/2$; $\alpha_{O1} = (4 + q - 2 \cdot r)/4$.

As $\gamma_S = (4 \cdot 1 + 1 \cdot q - 2 \cdot r)/1$, we obtain $\alpha_{O1} = \gamma_S / 4$.

The equation of energy metabolism gets the following form:

$$\alpha_{S1} \cdot CH_qO_r + \alpha_{O1} \cdot O_2 + ADP \rightarrow \alpha_{C1} \cdot CO_2 + \alpha_{H1} \cdot H_2O + ATP \quad (1)$$

Where: $\alpha_{S1} = 1/E_S$, $\alpha_{O1} = \gamma_S/(4 \cdot E_S)$, $\alpha_{C1} = 1/E_S$, $\alpha_{H1} = q/(2 \cdot E_S)$.

The plastic exchange under absence of essential amount of growth factors consumption can be described by the equation:

$$\begin{aligned} &\alpha_{S2} \cdot CH_qO_r + \alpha_{O2} \cdot O_2 + \alpha_{N2} \cdot NH_3 + E_X \cdot ATP \rightarrow \\ &CH_bN_cO_d + \alpha_{H2} \cdot H_2O + E_X \cdot ADP \end{aligned} \quad (2)$$

Let's express consumption of oxygen via a difference of degrees of initial and final products disoxidation.

For this purpose we shall write down the elementwise balance equations on carbon, hydrogen, oxygen and nitrogen:

$$\alpha_{S2} = 1; \quad \alpha_{S2} \cdot q + 3 \cdot \alpha_{N2} = b + 2 \cdot \alpha_{H2}; \quad \alpha_{S2} \cdot r + 2 \cdot \alpha_{O2} = d + \alpha_{H2}; \quad \alpha_{N2} = c$$

Then: $q + 3 \cdot \alpha_{N2} = b + 2 \cdot \alpha_{H2}$, substituting $\alpha_{N2} = c$, we obtain $\alpha_{H2} = (q + 3 \cdot c - b)/2$ and further from balance of oxygen atoms number follows:

$$\alpha_{O2} = [d + (q + 3 \cdot c - b)/2 - r]/2 = (q - 2 \cdot r + 2 \cdot d + 3 \cdot c - b)/4.$$

As $\gamma_S - \gamma_X = (4 + q - 2 \cdot r) - (4 + b - 3 \cdot c - 2 \cdot d) = q - 2 \cdot r + 2 \cdot d + 3 \cdot c - b$, then $\alpha_{O2} = (\gamma_S - \gamma_X)/4$.

Biomass synthesis (constructional exchange) can be described by energy coupling of the equation of formation ATP (1) and the equation of consumption ATP (2). We shall multiply all members of the equation (1) by E_X and summarize obtained equation with the equation (2), having substituted values of stoichiometrical coefficients:

$$\begin{aligned} &(E_X/E_S + 1) \cdot CH_qO_r + [E_X \cdot \gamma_S/(4 \cdot E_S) + (\gamma_S - \gamma_X)/4] \cdot O_2 + c \cdot NH_3 \rightarrow \\ &CH_bN_cO_d + (E_X/E_S) \cdot CO_2 + [E_X \cdot q/(2 \cdot E_S) + (q + 3 \cdot c - b)/2] \cdot H_2O \end{aligned} \quad (3)$$

Consumption of substrate – a source of carbon and energy – can be presented as the sum of expenditure for sustenance of vital functions (R_{Sn}), for synthesis of a cellular body in a plastic exchange (R_{Sx}) and for maintenance with supplementary energy of a plastic exchange (R_{Sg}). Thus according to the equation (3) we can write down consumption of substrate in a constructional exchange $R_{Sk} = R_{Sx} + R_{Sg}$. According to the equation (1): $R_{Sn} = \alpha_{S1} \cdot e \cdot X \cdot M_S/M_X = e \cdot X \cdot M_S/(M_X \cdot E_S)$, where e – specific speed of consumption of energy rich compounds for maintenance of ability to live, M_i – mass of a g-mole of the corresponding component.

According to the equation (3): $R_{Sk} = (E_X / E_S + 1) \cdot \mu \cdot X \cdot M_S / M_X$. In a result the following equation for metabolic speed of consumption of substrate is obtained:

$$R_S = \left[\frac{e}{E_S} + \left(1 + \frac{E_X}{E_S} \right) \cdot \mu \right] \cdot \frac{M_S}{M_X} \cdot X \tag{4}$$

Consumption of oxygen, in a similar manner to consumption of substrate, is also possible to present as the sum of consumption on distinguished metabolic blocks:

$$R_O = \left\{ \frac{e \cdot \gamma_S}{E_S} + \left[\gamma_S \cdot \left(1 + \frac{E_X}{E_S} \right) - \gamma_X \right] \cdot \mu \right\} \cdot \frac{M_O}{4 \cdot M_X} \cdot X \tag{5}$$

The carbondioxide production speed can be presented as production during consumption of substrate for maintenance of vital functions and during plastic exchange maintenance with supplementary energy:

$$R_C = (e + E_X \cdot \mu) \cdot \frac{M_C}{E_S \cdot M_X} \cdot X \tag{6}$$

Consumption of ammonia is determined according to the equation (2):

$$R_N = \alpha_{N2} \cdot \mu \cdot X \cdot M_N / M_X = c \cdot \mu \cdot X \cdot M_N / M_X \tag{7}$$

Biomass accumulation speed is determined by the formula:

$$R_X = dX / dt = \mu \cdot X \tag{8}$$

Thus, metabolic speeds can be expressed through the parameters which are material and energy characteristics of the cell growth process: μ, e, E_X.

The energy spent on synthesis of a biomass, cannot be less than consumption of energy rich compounds on synthesis of one gram of a biomass from given monomers (amino acids, purine and pyrimidine bases). Energy consumption in this case is by definition « a biological constant », which is equal about 10,5 g absolutely dried biomass / mole ATP [3, 4]. In our case it results in theoretical value E_X^O = 2,357 mole ATP / mole absolutely dried biomass. If γ_S is more than γ_X, then energy of the substrate, consumed in a plastic exchange, partially covers energy consumption for synthesis of a biomass, and the supplementary requirement for energy E_X can appear even less than E_X^O.

Experimental definition of composition of a biomass, values of any two metabolic rates and specific growth rate allows, using the equations (4) – (8), to find unknown parameters $(e, \ E_X)$ and to construct a full stoichiometrical pattern of a microbial biomass growing process.

Consumption coefficient (coefficient of an output) can be represented as ratio of metabolic rates of consumption of feeding components (or production of metabolites and products) to a biomass accumulation speed: $Y_{i/x} = R_i / R_X$, or in molar expression:

$$Y_{i/x}^{M} = \frac{M_X \cdot R_i}{M_i \cdot R_X} \text{ [mol i / mol X]} \tag{9}$$

The formula for calculation of a respiratory coefficient we shall obtain, using expressions (5) and (6):

$$RQ = \frac{R_C}{R_O} = \frac{4 \cdot (e + E_X \cdot \mu) \cdot M_C}{E_S \cdot \left\{ \frac{e \cdot \gamma_S}{E_S} + \left[\gamma_S \cdot \left(1 + \frac{E_X}{E_S} \right) - \gamma_X \right] \cdot \mu \right\} \cdot M_O} \tag{10}$$

Whence:

$$e / \mu + E_X = \frac{E_S \cdot (\gamma_S - \gamma_X)}{4 \cdot M_C / (M_O \cdot RQ) - \gamma_S} = F(1/\mu) \tag{11}$$

Obtained function $F(1/\mu)$ is linear with respects to $1/\mu$. Unknown values of energy rich compounds consumption on maintenance of ability to live (e) and the processes related to the growth (E_X), can be found from the graph drawn according to experimental data. For example, for the investigated strain of ethanol oxidizing yeast Candida lambica (the formula of biomass $CH_{1,807}N_{0,166}O_{0,44}$; $\gamma_X = 4{,}4282$; molecular mass taking into account ash elements M_X = 24,74) the corresponding results are represented on Fig. 1. As follows from the graph continued up to an axis of ordinates, E_X = 2,37 mole ATP/mole X, e is equal to tangent of angle of inclination a line of the graph to an axis of abscissas and is 1,683 mole ATP / mole X · hour (the data were obtained with the use of technique of least squares). The zero point of the above-stated dependence physically is not materialized, since it is achieved at indefinitely large value of specific growth rate.

The equations similar to (11) can be obtained also for others pairs of metabolic rates if instead of expression for RQ (10) other ratios are considered, for example, R_S/R_O. Sequence of operations for determination of unknown parameters will not be changed in this case.

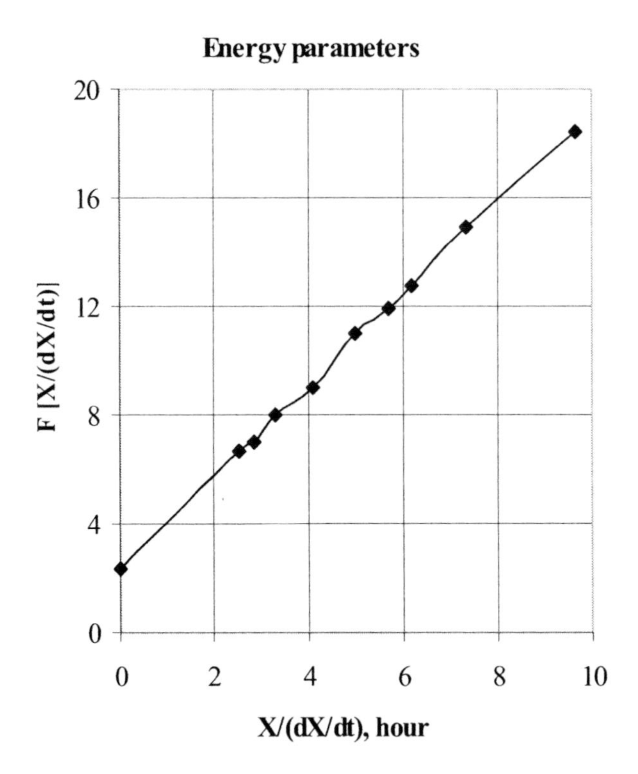

Fig. 1.

Let's designate enthalpies of combustion of the substrate and the biomass through H_S and H_X accordingly. Taking into account, that enthalpies of formation of carbon dioxide, water and ammonia are accordingly equal to h_C = 394 kilojoule/mole, h_H = 286 kilojoule/mole, h_N = 46 kilojoule/mole [5], and enthalpy of ATP hydrolysis is equal approximately h_F = 20,95 kilojoule/mole [6], according to the Guess's law enthalpy of formation of biomass and substrate are equal:

$$h_X = h_C + [(d - 3 \cdot c)/2] \cdot h_H + c \cdot h_N - H_X ;\ h_S = h_C + (q/2) \cdot h_H - H_S$$

If ethanol is used as a substrate (formula counted to one atom of carbon: $CH_3O_{0,5}$), then h_S = 145,6 kilojoule/mole. Owing to the known composition of a biomass h_X = 88,74 kilojoule/mole is obtained.

Taking into consideration the fact that in a plastic exchange the part of the substrate energy, proportional to a difference of degrees substrate and a biomass disoxidation, passes to a biomass, we determine additionally required consumption of ATP in a constructional exchange:

$$E_X^{(C)} = \frac{h_X - (\gamma_S - \gamma_X) \cdot h_S / \gamma_S}{h_{ATP}} = 2{,}415 \text{ mol ATP/mol X} \qquad (12)$$

The discrepancy between experimental value E_X and calculated value $E_X^{(C)}$ is: δ_E = 100 · |2,37 – 2,415| / 2,37 = 1,92 %. Inaccuracy first of all is determined by the fact, that the energy coefficient of efficiency (η_X) of a biomass synthesis process (plastic exchange) is a little bit less than 1 and is within the limits of 0,92 ÷ 0,97. Therefore calculated value $E_X^{(C)}$ can be used with adequate accuracy for drawing of the graph of linear dependence $F(1/\mu)$ by single definition of RQ and μ. The tangent of angle of lean the drawn linear relation is an assessment of rate of consumption of energy rich compounds in a metabolism of live maintenance.

After determination values of parameters e and E_X, all unknown metabolic rates can be calculated (for any predetermined value of specific growth rate). Their calculation in dimension mol / mol X · hour directly gives values of stoichiometrical coefficients in the summarized equation of the material balance. The stoichiometrical coefficient at a sign of molecule of water in the summarized equation of stoichiometry microbial synthesis is calculated by balance of number of hydrogen atoms.

Operative measurement of only two metabolic speeds allows to calculate stoichiometry of the growth process in "on-line" mode. For calculation of aerobic processes it is preferable to use gas analysis. In this case not only the data for definition R_O and R_C on the basis of a gas stream balance are provided, but also calculation of specific growth rate value is carried out on time intervals on which there were no process interventions, changing speed of a gas stream or volume of a liquid phase (single samplings, essential filling-up of medium, dilutions, etc. operations). For rather short time intervals it is possible to assume $\mu = \text{const}$, and to write down the following expressions for the ends of an interval (for the moments of time $t_{(i)}$ and $t_{(i+1)}$): $R_{O(i)} = (\alpha_O + \beta_O \cdot \mu) \cdot X_{(i)}$; $R_{O(i+1)} = (\alpha_O + \beta_O \cdot \mu) \cdot X_{(i+1)}$.

Taking the logarithm of these expressions, subtracting the first from second and dividing both parts of the obtained equation on $\Delta t = t_{(i+1)} - t_{(i)}$, we obtain:

$$(\Delta \ln X)/\Delta t = (\Delta \ln R_O)/\Delta t \text{ ; and } \mu \approx (\Delta \ln R_O)/\Delta t \qquad (13)$$

If the essential discrepancy between experimental and calculated values of E_X takes place, or a deflection from linearity of the dependence represented on Fig. 1, then it means that by-products are likely to be formed (for example, during cultivation of ethanol oxidizing yeast in conditions of a comparative limit on oxygen, acetic acid is formed). In this case the stoichiometrical equations must be supplemented with the equation of synthesis (and possible secondary consumption) of exometabolite.

On the Fig. 2 dependence of economic coefficient on the time of consumption of a substratum dose is shown. Process was carried out in a mode of oxystate. The pronounced extreme character of this dependence is evidence of a change of ratios between metabolic processes at variable concentration of substrate. For an estimation of the possible mechanism of the process comparative experiments have been executed on utilization of ethanol and products of its successive oxidation – acetaldehyde and an acetic acid. The results represented in table 1, affirm that the acetic acid, formed at relative excess of ethanol, can be consumed by studying culture of yeast in conditions of a lack of substrate, but practically exclusively in energy metabolism. Thus on one conditional g-mol of a formed acetic acid (counting upon one atom of carbon), the cell obtains amount of energy E_P = 2,5 moles ATP, and on its consumption – amount $E_S - E_P$ = 5,5 moles ATP.

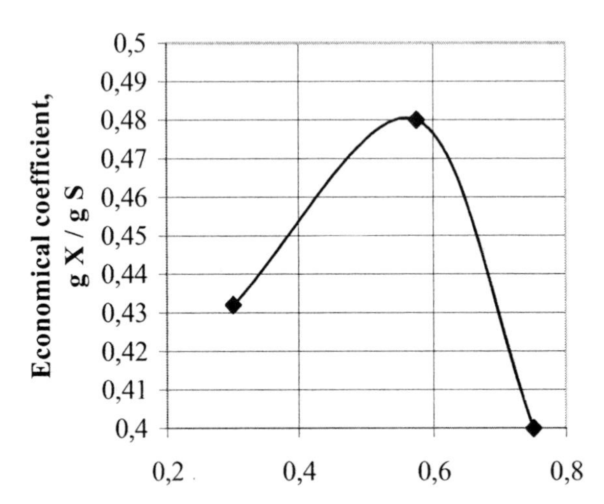

Fig. 2.

Table 1. Efficiency of consumption of products of consecutive ethanol oxidation

Substrate	Respiratory coefficient, g CO2/g O2	Economic coefficient, g X/g S
Ethanol	0,77	0,49
Acetaldehyde	0,97	0,48
Ammonium acetate	1,37	0,08

*The stoichiometry of production an acetic acid t*aking into account its automatic titration by ammonium hydroxide is described by the equation:

$$\alpha_{S3} \cdot CH_qO_r + \alpha_{O3} \cdot O_2 + \alpha_{N3} \cdot NH_3 + ADP \rightarrow \alpha_{P3} \cdot CH_kN_nO_f + \alpha_{H3} \cdot H_2O + ATP \quad (14)$$

Where: $\alpha_{O3} = (\gamma_S - \gamma_P)/(4 \cdot E_P)$, $\alpha_{N3} = n / E_P$, $\alpha_{H3} = (\gamma_S - \gamma_P)/2 + r - f$,

$\alpha_{S3} = \alpha_{P3} = 1/E_P$.

Let's designate amounts of moles ATP formed on mechanisms of reactions (1) and (14), as y and z accordingly. Molar coefficients of consumption and synthesis of components can be represented as:

$$(M_X \cdot R_O)/(M_O \cdot R_X) = y \cdot \alpha_{O1} + \alpha_{O2} + z \cdot \alpha_{O3} \tag{15}$$

$$(M_X \cdot R_C)/(M_C \cdot R_X) = y \cdot \alpha_{C1} \tag{16}$$

$$(M_X \cdot R_N)/(M_N \cdot R_X) = \alpha_{N2} + z \cdot \alpha_{N3} \tag{17}$$

$$(M_X \cdot R_S)/(M_S \cdot R_X) = y \cdot \alpha_{S1} + \alpha_{S2} + z \cdot \alpha_{S3} \tag{18}$$

$$(M_X \cdot R_P)/(M_P \cdot R_X) = z \cdot \alpha_{P3} \tag{19}$$

From the equations (16) and (17) we shall express y and z, substitute in the equation (15) and resolve it with respect to rate of biomass growth:

$$\frac{1}{M_X} \cdot \frac{dX}{dt} = \frac{4 \cdot R_O / M_O - \gamma_S \cdot R_C / M_C - (\gamma_S - \gamma_P) \cdot R_N / (n \cdot M_N)}{\gamma_S - \gamma_X - (\gamma_S - \gamma_P) \cdot c / n} \tag{20}$$

Similarly with use of other equations of system (15) – (19) it is determined:

$$\frac{1}{M_S} \cdot \frac{dS}{dt} = \frac{4 \cdot R_O / M_O - \gamma_S \cdot R_C / M_C - (\gamma_S - \gamma_P) \cdot (dX/dt)/M_X}{\gamma_S - \gamma_P} \tag{21}$$

$$\frac{1}{M_P} \cdot \frac{dP}{dt} = \frac{4 \cdot R_O / M_O - \gamma_S \cdot R_C / M_C - (\gamma_S - \gamma_X) \cdot (dX/dt)/M_X}{\gamma_S - \gamma_P} \tag{22}$$

For a case of the secondary consumption of ammonium acetate in energy metabolism the equations completely identical to (20) – (22) are obtained. Thus, from six metabolic rates it is necessary to measure only three. The others are determined from stoichiometrical invariants (20) – (22).

Having expressions for metabolic speeds, it is possible to write down the equation of balance of energy rich compounds (for a condition of a constancy of their endocellular pool):

$$(y + z) \cdot (dX/dt)/M_X = E_X \cdot (dX/dt)/M_X + e \cdot X/M_X \tag{23}$$

Or, taking into consideration expressions y and z, obtained from the equations (16) and (17), the equation (23) can be written in the form:

$$\left(E_S - \frac{E_P \cdot \gamma_S}{\gamma_S - \gamma_P}\right) \cdot \frac{M_X \cdot R_C}{M_C \cdot (dX/dt)} + \frac{E_P}{\gamma_S - \gamma_P} \cdot \left[\frac{4 \cdot M_X \cdot R_O}{M_O \cdot (dX/dt)} - \gamma_S + \gamma_X\right] = E_X + e/\mu \tag{24}$$

The equation (24), in a similar manner to the equation (11), allows to find energy parameters E_X and e on the basis of the corresponding data of gas analysis and experimental values of a biomass concentration.

Diagnostics of a biosynthesis process on the basis of the equations such as (11) and (24) gives directly numerical values of parameters. However, use of the numerical data for the purposes of decision-making is not always convenient. For the operator, who is controlling the process, the information in the visual form is preferable. For example, it can be a phase plane on which the trajectory of system is displayed. Traditionally concentrations are used as phase coordinates. But, due to the nonlinearity of the equations of kinetics, stoichiometrically different zones on such planes will be separated by the nonlinear borders, determined by unknown kinetic parameters of system. We offered to use of specific metabolic rates, ratios between which are always linear due to linearity of the stoichiometrical equations, as phase coordinates. Such phase plane is a diagnostic plane and is characterized by linearity of the borders which are dividing stoichiometrically various zones. Process of a biosynthesis is displayed on a diagnostic plane as a projection of a trajectory of system by software of a workplace of the operator-technologist. For aerobic processes it is preferable to use the phase coordinates system R_O/X – R_C/X (Fig. 3), in which all variants of stoichiometry are visualized. The plane is separated by linear borders into 4 zones. The first zone is separated by a straight line with the tangent of angle of lean equal to a respiratory coefficient of process of full oxidation of the exometabolite – acetic acid (RQ_1 = 1,375 g CO_2/g O_2). The second border is a straight line corresponding to process of full oxidation of ethanol (RQ_2 = 0,917 g CO_2/g гO_2). The third line (dashed line) corresponds to value respiratory coefficient, calculated from the equation of the balanced growth. It is theoretically optimal mode in the absence of substrate consumption on maintenance of ability to live. According to the equation (3) $RQ_3 = 4 \cdot E_X / [\gamma_S \cdot E_X + (\gamma_S - \gamma_X) \cdot E_S]$ = 0,354 g CO_2/g O_2. The fourth border corresponding to production of exometabolite coincides with an abscissa axis ($RQ_4 = 0$). All lines of borders, except for the third, pass through the origin of coordinates. The third begins from a point laying on a line of full substrate oxidation and having coordinates, corresponding to consumption of substrate for maintenance of vital functions:

$$\alpha_O = \alpha_{OI} \cdot e \cdot M_O / M_X = \gamma_S \cdot e \cdot M_O / (4 \cdot E_S \cdot M_X) = 0{,}408 \text{ g } O_2/\text{g X·h};$$

$$\alpha_C = \alpha_{CI} \cdot e \cdot M_C / M_X = e \cdot M_C / (E_S \cdot M_X) = 0{,}374 \text{ g } CO_2/\text{g X·h}.$$

Angle of lean of the third border corresponds to value RQ_3.

The lower border of the third area is described by the linear equation which we shall name the equation of the balanced growth: $R_C / X = (R_O / X - \alpha_O) \cdot RQ_3 + \alpha_C$.

Diagnostic planes can be constructed similarly in coordinates R_O/X – R_S/X; R_S/X – μ, etc. Set of such graphic information formed on the display in a real time mode, allows the explorer or the operator-technologist to estimate immediately a state of process.

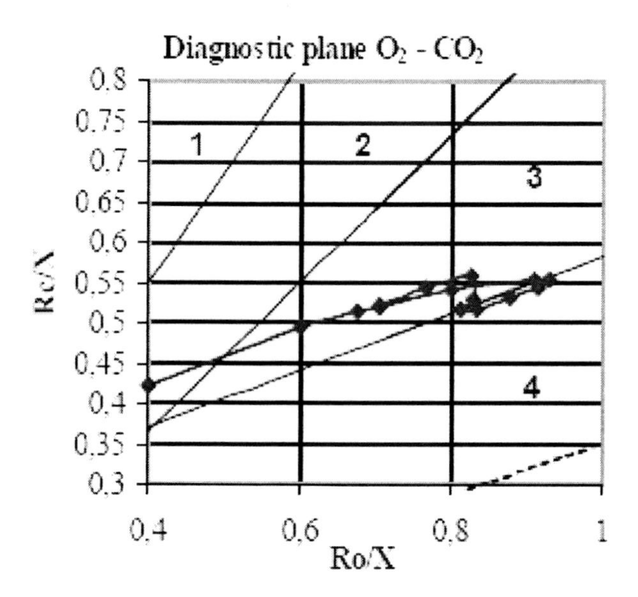

Fig. 3.

References

[1] S.J. Pirt: *Proc. Roy. Soc.*, 224 (163), (1965).

[2] G.Enikeev, S.G.Muhachev, R.I.Valeev: The Soviet-Finnish seminar on biotechnology. Moscow, *Glavmicrobioprom*, 1983. (in Russian).

[3] J.C. Senez: *Bact. Rev.*, 95 (26), (1962).

[4] V.N.Ivanov: Energetics of microorganisms growth. *Naukova Dumka*, Kiev 1981. 140 p. (in Russian).

[5] The directory of the chemist. State scientific.-tech. publishing house of chem. *literature*. Moscow. Leningrad. (1), 1962.1072 p. (in Russian).

[6] N.M.Manakov: *Applied biochem. and microbiol.*, 375 (3), 17 (1981). (in Russian).

In: Industrial Application of Biotechnology
Editors: I. A. Krylov and G. E. Zaikov, pp. 85-89
ISBN 1-60021-039-2

Chapter 10

SELECTION OF MUTANTS IN THE 25-HYDROXYCHOLESTEROL RESISTANCE SYSTEM UNDER THE INFLUENCE OF N-NMU

S. L. Karanova*, V. A. Paseshnichenko,[1] I. E. Kulichenko, I. E. Knyaz'kov, A. M. Nosov, N. V. Proliotova[2], N .A. Bogoslovskii[3] and R. G. Kostyanovskii[4]

Tymiryazev Plant Physiology Institute RAS, Moscow, Russia
[1]Bach Biochemical Institute RAS, Moscow, Russia
[2]All-Russian Flax Rasearch Institute, Torzhok, Tver region, Russia
[3]Vitamin Institute RAS, Moscow, Russia
[4]Physical Chemistry Institute RAS, Moscow, Russia

ABSTRACT

The selective system is offered to reveal the mutant plant cells characterized by the changed isoprenoid biosynthesis regulation at the level of HMG-CoA reductase. 25-Hydroxycholesterol, the effective inhibitor of HMG-CoA reductase, synthesized in Vitamin Institute RAS (Moscow), was used as a selective agent. To increase the frequency of genetic changes, high-effective chemical mutagen N-nitroso-N-methylurea (N-NMU) was used at the dose of 0,5 mM·h.
These conditions of cell selection were approved by using *Dioscorea deltoidea* cell suspension (dedifferentiated cell culture) and several lines of morphogenic long-fibre flax callus culture. The repeated treatment of *Dioscorea deltoidea* cell strain PPI DM0,5 (characterized by furostanolic saponin superproductivity) was also carried out using the some dose (0,5 mM·h) of N-NMU.
Irrespective of provided selective conditions, the frequency of genetic changes, which determined *Dioscorea deltoidea* cell resistance to 25-Hydroxycholesterol, was about 10^{-7} of plated alive cell number. During the period of active growth, we noticed colour variation (from light to greenish and yellowish) among Dioscorea cell clones. These

* Tymiryazev Plant Physiology Institute RAS, ul. Botanicheskaya, 35; 127276, Moscow, Russia; tel.:(095) 903 93 34; fax: (095) 977 80 18; e-mail: gsc@ippras.ru

selective conditions also very effectively allowed to reveal the mutant cells in control cell population.

Key Words: cell selection; HMG-CoA reductase; 25-Hydroxycholesterol; mutagenesis; N-NMU

Introduction

The use of experimental mutagenesis approach and cell selection on the level of somatic cells in vitro can provide an opportunity of obtaining biochemical mutants capable of giving a clue to understanding how isoprenoids biosynthesis is regulated in plant cell. Such mutants may have properties which appear to be of practical value. It has long been proved that as well as in animal tissues, HMG-CoA reductase is a key enzyme in the isoprenoid biosynthesis in plants (Bach, Lichtenthaller, 1982; Paseshnichenko, 1984; bach, 1995). We are not aware of any research to create specific selective conditions aimed at getting mutants with changed regulation of isoprenoid biosynthesis at HMG-CoA reductase level in plant cells. However, it is known that mevinolin and compactin, which are HMG CoA reductase inhibitors, proved to be powerful inhibitors of phytosterol biosynthesis in radish sprouts (Bach, Lichtenthaller, 1982) and platan tissue culture (Ryder, Goad, 1980). 25-Hydroxycholesterol is the most powerful and specific HMG-CoA reductase inhibitor among cholesterol analogeus (Schroepfer, 1981).

Our work was aimed at developing a selective system to reveal 25-Hydroxycholesterol resistant mutant cells from cell populations in vitro *Dioscorea deltoidea* Wall producent of steroids with possibly changed regulation of their biosynthesis as well *Linum usitassimum* L.- for perspective obtaining patogen resistant species with traits of practical value.

Materials and Methods

Biological material. The starting material for the research included:

- Suspension culture of *Dioscorea deltoidea* Wall cells, strain IPPDM0,5 without morphogenesis traits, superproducer of furastanol saponins with aglycon diosgenin wich was obtained earlier (Karanova et al., 1986; Butenko et al., 1987) after treatment the original strain IPPD1, wich had been obtained by Butenko R.G. and Abroshnikova M.A. (Sarkisova, 1973) with a growth stimulating dose of an effective chemical N-NMU mutagen (Karanova et al., 1973).
- Flax callus culture with morphogenesis traits of several genotypes (*Linum usitatissimum* L.) taken from the collection of the Biotechnology laboratory (ARFRI, Torzhok) (Proletova, 2003).

Selective system. The selective agent used was an effective inhibitor of key enzyme in isoprenoid biosynthesis HMG CoA reductase 25-Hydrohycholesterol wich had been synthesized in Vitamin Institute RAS (Moscow). To prepare the stock solution the necessary quantity of the analogue was first dissolved in a small amount of 96% ethanol, then supplement with the propper liquid nutrient medium. Sterilization was effected through membrane filters.

Treatment with mutagen. The research made use of a highly effective chemical N-NMU mutagen in liquid nutrient medium with pH=5,6. The mutagen had been synthesized in the Institute of Chemical Physics RAS (Moscow). The stock mutagen solution was sterilized by passing through a glass Schott filter and then added aseptically to cell suspension. The cells were exposed to the mutagen for 1 h and then washed 5 times with excess fresh nutrient medium and then placed in nonselective and selective nutrient agar medium (5 test-tubes per one variant of long-fibre flax) or were placed into flasks with liquid nutrient medium and kept there for 7 days for phenotypical expression after wich the cell suspension was mixed with nonselective and selective medium and poured into Petri dishes (10 dishes for one variant in the case of *Dioscorea deltoidea*).

The cell suspension density (the number of cells per 1 ml) was measured in Fuchs-Rosenthal hemocytometer.The cultures had been macerated in 20% chromic acid for 10-20 min at 70°C. The density (x) was calculated according to the formula: x=1000xM/3.2, with M being the average number of cells per compartment with 6 repetitions.

The cell viability (% of living cells) was evaluated by the method of vital staining with 0.1% Evans blue. No less than 200 cultured units were analysed in each variant.

Cloning was performed according to the method proposed by Bergmann (Bergmann, 1960). The prepared cell suspension wich contained both free cells and small aggregates with the density 10^5 cell/ml was mixed the equal volume of nutrient medium containing 1.6% of agar at 40°C and poured in plastic Petri dishes (10 ml per a 9-cm diameter dish or 2 ml per 4-cm diameter dish). The development of the colony was watched for 4 weeks. For futher cloning only viable colonies were selected, as we had already determined that a cell should undergo no less than 9 divisions in order to form a viable clone (Karanova, 1977). After 4 weeks of cultivation the number of viable clones in nonselective and selective media was calculated.

RESULTS

1. *Characterization of the selective system.* The preliminary experiments had selected an effective analogue concentration. *Dioscorea deltoidea* cells were plated on nutrient media with 25-Hydroxycholesterol concentrations: 0.025, 0.05, 0.075 and 0.1 mcM. Colonies of different sizes were formid after 3-4 weeks cultivation. 25-Hydroxycholesterol in concentration 0.1 mcM exerted lethal effect. Only microcolonies wich failed to survive were formed. Inhibition of colony formation was observed at 0.075 mcM concentration of 25-Hydrohycholesterol. In later experiments we used selective media with analogue concentration 0.025, 0.05, 0.075 mcM for *Dioscorea deltoidea* and 0.05 mcM for long-fibre flax.
2. *Selection of Dioscorea deltoidea clones resistant to 25-Hydroxycholesterol.* The most detailed description of resistent clones selection process was made for *Dioscorea deltoidea*. Initially, in the 1st passage, the total number of the clones obtained at different analogue concentrations after N-NMU treatment and without it

was about 400. However, most of them failed to survive even in the second passage. All viable clones were tested for growth in the absence of selective conditions. Besides, their growth at different 25-Hydroxycholesterol concentrations was studed. The selection effectiveness depended on the experimental variant. When the cells which were not treated with N-NMU were plated on the medium with 0.025 mcM analogue concentration, the portion of stable clones was about 3%, and it was about 9% at 0.05 mcM. After repeated treatment with 0.5 mMxh dose of N-NMU and plating on 0.05 mcM 25-Hydroxycholesterol the portion of viable clones was more than 14%. In the meantime it was noted that single large colonies formation on selective medium and their growth stimulation occurred 2 weeks after culturing while without N-NMU treatment it happened only 4 weeks later. This supports our previous observations made on plant-producent cell cultures of biologically active compounds *Dioscorea deltoidea, Yucca gloriosa*, *Ajuga turkestanika*, *Medicago sativa, Strophanthus gratus* wich were once treated with N-NMU in the range of small doses 0.1-2.0 mMxh (Karanova, 1999).

Discussion

The results obtained enable us to state that repeated treatment of *Dioscorea deltoidea* cells with a small N-NMU dose (0.5mMxh) has revealed inherited changes induction which provides stable growth of the cells on 25-Hydroxycholesterol. The frequency of genetically chanched cells occurencence was about 10^{-7} per the number pf plated cells. In the second and follwing passages a variety of *Dioscorea deltoidea* clones was revealed whose color ranged from light to greenish and yellowish. The yellowish color of clones was intensified after their exposure to light. Thus we have suggested effective selective conditions to reveal the cells resistent to 25-Hydroxycholesterol wich is an inhibitor of a key enzyme in isoprenoid biosynthesis. Earlier, it was shown on animal cells that 25-Hydroxycholesterol in the concentration range used in our experiments inhibited HMG-CoA reductase and sterol synthesis (Kandutsch, Chen, 1974).

References

[1] T.J.Bach, H.K.Lichtenthaller Z. *Naturforsch*. 37c:46 (1982).

[2] V.A.Paseshnichenko Sterol saponins of Dioscorea-type plants: structure, metabolism, biological activity. *Doctoral (Biol.) Dissetation*, *IBch* (1984) (in Rissian).

[3] T.J.Bach *Lipids* 30: p:191 (1995).

[4] N.C.Ryder, L.J.Goad Biochemica et *Biophysica Acta* 619: p:424 (1980).

[5] S.L.Karanova, A.M.Nosov, V.N.Paukov, Z.B.Shamina Plant Cell Culture and Biotechnology, Butenko R.G.,Ed., Moscow: *Nauka*, p:83 (1986) (in Rissian).

[6] R.G.Butenko, S.L.Karanova, Z.B.Shamina, A.M.Nosov USSR Inventor's Certificate no. 1389283, C12, N5/00 (1987) (in Rissian).

[7] M.A.Sarkisova *Dioscorea deltoidea* Wall tissue culture as a sterol saponins producent. *Cand. Sc. (Biol.) Dissertation*, Moscow, IPP (1973) (in Rissian).

[8] S.L.Karanova, R.Dalenburg, Z.B.Shamina, S.I.Demchenko, I.A.Rapoport Application of Chemical Mutagens in Agriculture and Medicine. Rapoport I.A., Ed., Moscow: *Nauka*, p:130 (1973) (in Rissian).

[9] N.V.Proletova Anther culture in the selection of long-fibre flax resistant to fusarium wilt *Cand. Sc. (Biol.) Dissertation*, Moscow, *MAA* (2003) (in Rissian).

[10] l.Bergmann *J. Gen. Physiol.* 43: p:841 (1960).

[11] S.L.Karanova Induced Mutagenesis in the Culture of Somatic Cells of *Dioscorea deltoidea* Wall. *Cand. Sc. (Biol.) Dissetation*, Moscow, IDB (1977) (in Rissian).

[12] S.L.Karanova Principles of obtaining practically valuable strains of plant cell cultures by experimental mutagenesis method. *Doctoral (Biol.) Dissetation*, Moscow, *RChTU* (in Rissian) (1999).

[13] A.A.Kandutsch, H.W.*Chen J. Biol. Chem.* 249: p:6057 (1974).

In: Industrial Application of Biotechnology
Editors: I. A. Krylov and G. E. Zaikov, pp. 91-95
ISBN 1-60021-039-2

Chapter 11

BIOSORBENTS FOR RECEPTION OF BIOCATALYSTS OF MEDICAL PURPOSE

S. M. Kunizhev*, O. V. Vorobyeva, A. A. Fill and O. V. Anisenko
Stavropol State University, Stavropol

ABSTRACT

The methods for immobilization of the urease on biosorbents obtained by surface activation of pyrogenic silicon dioxide – aerosil and microcrystallic cellulose using of an albuminous complex of a casein. The catalytically stable and active ferment preparations were obtained. The effect of pH and temperature on activity of both dissolved and immobilized ureases was investigated.

Key words: biosorbent, immobilization of the urease, microcrystallic cellulose, pyrogenic silicon dioxide – aerosil.

Creation natural catalysts on the basis of immobilized enzymes and their use in the medical purposes, in bioreactors and numerous analytical devices is one of the important problems of scientific researches. The advantage them before natural enzymes consists in greater stability and an opportunity branches immobilized enzyme from reaction product. In this connection use of qualitatively new sorption materials for immobilization enzymes will help with the decision of one of the important problems of modern biotechnology.

The purpose of the given work was study of processes immobilization urease on the biosorbents received by a method of an activation of a surface pyrogenic silicon dioxide – aerosil and microcrystallic cellulose (MCC) of casein, and also research of influence of various factors on properties soluble and immobilized enzyme. The urgency of the given problem is caused by reception preparations immobilized enzyme with high percent preservation activity.

---

* Stavropol State University, 1, Pushkina St., Stavropol, 355009 ; Tel. 7(8652)35-91-34, Fax: 7(8652)35-40-33; E-Mail: Biochem@Stavsu.Ru

Immobilized preparations urease important as the test-systems at definition of a level of urea in blood and plasma.

Experiment

For immobilization of urease used the enzyme allocated from soybeans on a method, described Meshcovoi. [1] with specific activity 34,60 mcmol /mg of enzyme.

Activity immobilized and natural enzyme estimated spectrofotometric method where as a substrate used of 3 % of urea solution in the phosphatic buffer pH 7,0 [2].

Modelling and synthesis of sorbents on the basis of aerosil, microcrystallic cellulose and casein carried out according to techniques [3, 4].

Immobilization of urease spent was 24 hours at temperature 4 °C as follows: to one part of a damp biosorbents added double volume of a solution of enzyme in the phosphatic buffer pH 7,0 with concentration urease 4,0 mg/ml. Not contacted with the carrier a urease deleted washing of a matrix by the distilled water before reception of negative test on presence of fiber.

Quantity immobilized urease defined on a difference concentration enzyme in an initial solution and in a solution after immobilization on method Louri [5.] Immobilized a preparation stored at temperature + 4°C.

Results and Discussion

By means of the chosen methods immobilization urease were obtained of preparations with a high level of preservation of specific activity which expressed in percentage of specific activity of soluble enzyme (table).

Table. The performance of immobilized drugs of a urease

Carrier	Specific activity of enzyme urease, mcmol/mg of enzyme	Preservation of activity immobilized urease, %	Quantity immobilized enzyme (on fiber)	
			mg/g of a sorbent	% from the maintenance of enzyme in a solution
MCC + casein	33,91	98	5,60	70
Aerosil + casein	34,60	100	5,04	63

For definition of value pH at which the greatest was observed enzyme activity soluble and immobilized urease, use solutions of the phosphatic buffer with pH 5,4; 6,0; 7,0; 8,0.

For an establishment of a range termostability soluble and immobilized urease phosphatic buffer solutions with pH at which the greatest activity of enzyme was observed have been used. Samples soluble and immobilized urease maintained in an interval of temperatures from 10 up to 70°C within 24 hours.

With the purpose of optimization of intervals of a variation of the factors influencing activity soluble and immobilized urease, the analysis a contour of a surface with use of program Statistic Neural Networks. As bases of the factors influencing on catalytically activity immobilized of enzyme, have been chosen: temperature of carrying out enzyme reactions and pH environments. For definition of boundary conditions of operating factors data from appendix Neural Networks imported to a package of applied programs Statistic. In figure 1 the surface of the response between two factors (temperature and pH environments) and function as which specific activity immobilized enzyme acts is represented.

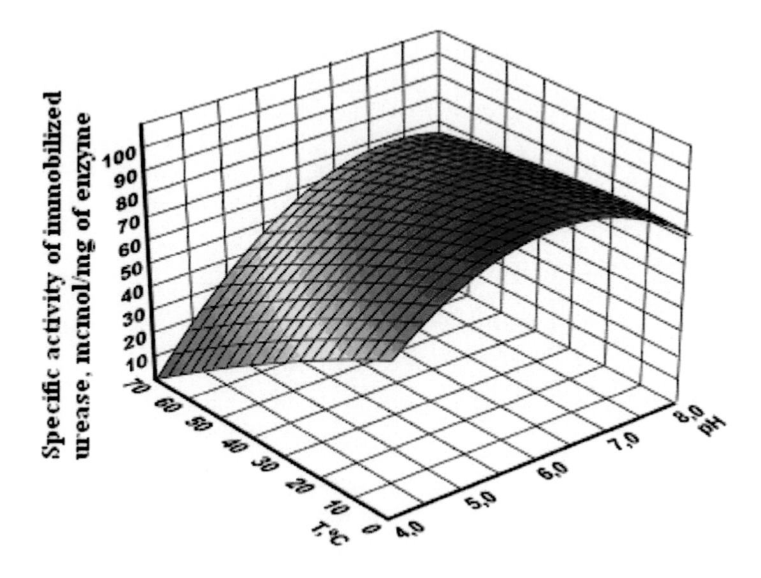

Fig. 1. Optimization of factors (temperature and pH environments), influencing specific activity immobilized urease

As a result of the analysis of interfactorial interactions are revealed option predicted factors: temperature - 20÷30°C, pH environments - 6,5÷7,0. The data received at planning of experiment, prove to be true resultats experiment.

Influence pH environments on activity soluble and immobilized urease presented in figure 2. For soluble urease the optimum of specific activity was observed at pH 7,0. Urease immobilized on aerosil with casein, had an optimum of specific activity at pH 6,0; and on MCC with casein 7,0. Values of specific activity accordingly made 34,60 and 33,91 mcmol/mg of enzyme.

Obtained data about influence of temperature on activity natural and immobilized urease are presented in figure 3.

Soluble urease termostability at 20 ^{0}C. For immobilized urease the range termostability lays within the limits of 10÷70 ^{0}C.

Stability of immobilized urease estimated on residual activity after storage at temperature +4°C within three months. Thus residual activity was measured periodically with 2-3 times a week. Soluble urease reduces activity after two weeks storage at temperature +4°C on 80 %. Preparations immobilized urease kept initial activity within three months. Preservation of specific activity of enzyme, immobilized on the carrier, at multi-fold use allows to save considerably expenses at obtained a fermental preparation.

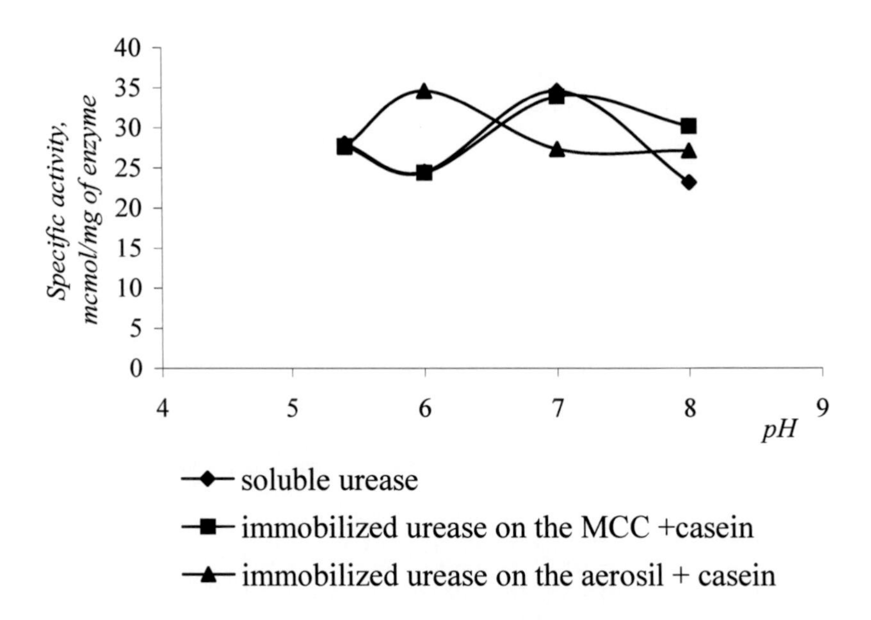

Fig. 2. Influence of value pH on specific activity soluble and immobilized urease

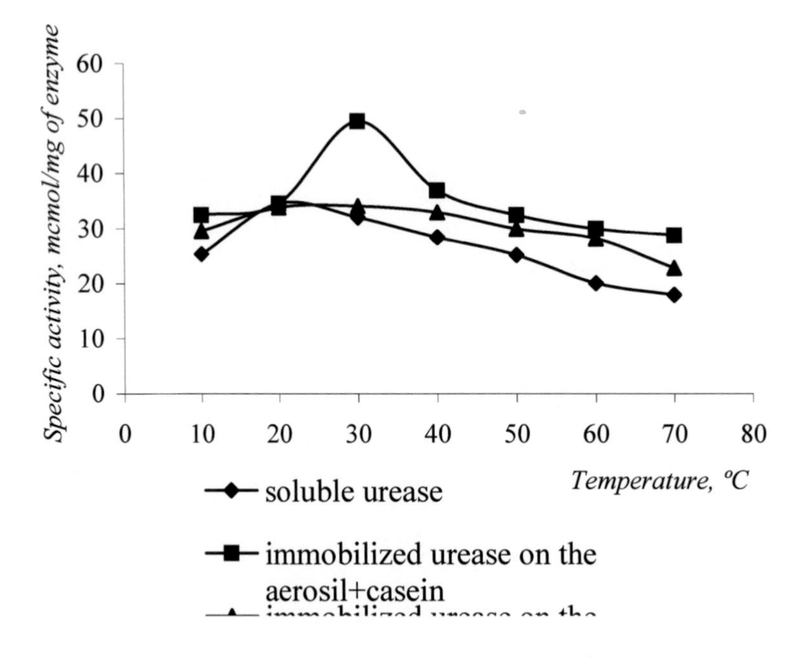

Fig. 3. Influence of temperature on specific activity soluble and immobilized urease

Advantages of immobilized preparations urease, characterized termostability, an opportunity of repeated use, as a whole allow to raise efficiency of practical application use in various areas of a science and manufacture.

REFERENCES

[1] Meshkova N.P., Severin S.E. *Practical work on biochemistry*. M.: Publishing house of the Moscow State University, 1979. - P. 430.
[2] Cochetov G.A. *Practical guidance on enzymology*. Moscow, 1980. – 43p.
[3] Kunizhev S.M., Vorobyova O.V. New sorption materials for biotechnology// *Scientific Israel – technological advantages*. - 2004. - Vol 6, № 1-2. – P. 15 –18.
[4] Ishchenko V.M., Vorobyeva O.V. Mathematical of model in biotechnologysorbents // *Russian Academy of Science*, 2004. - P.191 - 193.
[5] Menshikova O.V. Laboratory methods research in clinic. - M.: *Medicine,* 1987. - P. 26-27.

In: Industrial Application of Biotechnology
Editors: I. A. Krylov and G. E. Zaikov, pp. 97-104
ISBN 1-60021-039-2

Chapter 12

ECOBIOTECHNOLOGICAL METHODS OF PROCESSING FUR RAW MATERIALS

Dm. V. Shalbouyev* and V. S. Dumnov
East-Siberia State University of Technology,Ulan-Ude

Characteristics of sewage waters forming under treatment of different types of raw materials were showed. The influence of synthetic surface active substances (SSAS) of different chemical nature on bacterial suspensions properties were investigate. Ecobiotechnological degreasing process allow to reduce the level of toxic contamination of sewage waters and provide optional elimination of greases from hair and skin tissues of sheepskin.

Bacterial suspension, greasing substances, synthetic surface active substances, Pseudomonas sp, sewage water, hair and skin tissues.

Traditional methods of sheepskin treatment are based on the water solutions containing various chemical substances with highly toxical properties. The most contaminated sewage waters are formed after the emulsion degreasing of sheepskin materials. The chemical analysis of sewage waters has revealed that they contain a great number of suspended and greased substances, synthetic surface active substances (SSAS) up to 6,0-7,5 g/l, formaldehyde (up to 0,5 g/l) and possess poor alkaline reaction (pH is about 7,5-8,0) which allows to characterize them as hypertoxic or very contaminated.

The application of biotechnological principles of improving the process of degreasing is caused by the following requirements:

(1) The process of microbiological degreasing allows to reduce the level of technogenic influence on the environment through exclusion of formaldehyde from degreasing combination, sodium carbonate and a considerable reduction SSAS from 8,0 to 0,5 g/l.
(2) The applied composition for biotechnological degreasing should possess both lipase and protease activities, which help to destruct, grease in skin and hair materials.

(3) The quantity of protease activity should be minimum because it allows to destruct protein surface of grease substances on sheepskin without loosening hair tension.

When estimating the modern enterprise activity it is necessary to take into account not only its production efficiency but also the expected level of technogenic effect on the environment not excluding the fur treating production enterprises [1,2]. The sewage waters being formed after the sheepskin fur treatment present a heterogenic system including pollutants of different chemical nature [3]. It should be pointed out that the composition of sewage waters of certain enterprises may sufficiently differ in concentration and composition of contaminating substances. This sufficient differentiation of qualitative characteristics of sewage waters is determined by the sort of the processed raw materials, the chemical materials and the technological map of fur production materials.

The comparative characteristics of sewage waters after the treatment of different sorts of raw material revealed considerable degree of contaminated sewage waters in the case of sheepskin treatment. In this case of treatment sewage waters are characterized by high level of contamination with both easily and difficult oxidizing organic substances which is proved by the maximum values of BOD and COD. The microstructure of skin material to the greatest degree determines the qualitative composition of sewage waters after the process of pickling. Skins with thin and friable middle skin are treated in poorer pickling solutions than those with thick and douse middle skins which need to be treated with much stronger pickling. On comparing the amount of acidity of the used pickling solution it can be seen that under the growing density of collagen fibres in the middle skin structure of sheepskin and deer paws and bear skin as well the acidity index grows from 2,5 up to 10,4 and 15,9 g/l with the lowering of active reaction in the environment from 2,8 to 2,3 and 1,5 respectively. After the treatment of fur materials only the amount of acidity of pickling solution drops out from the given number, which makes up 5,3 g/l and considerably prevails the acidity index of sewage water after the sheepskin treatment. It is evidently connected with the treatment of the skins with long and thick hair surface like sheepskin fur which absorbs the greater portion of acid during the process. Besides, the organic acid is usually absorbed in much less quantities than in comparison with sulphuric acid.

For the prevention of acid tumour of middle skin the treatment of raw material is carried out in the acid and salt solution, containing neutral salt along with the acids, with the portion not lower than 40 g/l. The average degree of sodium chloride after pickling constitutes not more than 10% which explains the high concentration of this ingredient in sewage water after the process of pickling.

The estimation of the level of toxic contaminants in sewage waters, forming after the treatment of fur material allows to determine the degree of negative effect produced by industrial enterprises on the environment. The level of toxic contaminants allows to give the total information about the quality of sewage waters coming both on the purification units and discharging into the native water objects. For the accomplishment of this aim it is possible to apply the toxicological control using the methods of biotesting [4, 5]. The following organisms – Daphnia magna Straus were used as test-objects. The evidence of the degree of toxic substances in the processing solutions and sewage waters was the percent of the survival of the Daphnia magna Straus. For the control and diluting of sewage waters ordinary water from water supply system was used. Observations were being carried out during 96 hours; during the first three hours the probes were taken hourly, then once in the period of twenty four hours the number of the survived samples and their reaction was being fixed. On completion exposure the level of toxic contamination of the waters under control was determined. After the 96 hours observation survival constituted 100% where as the maximum survival in the solutions under control were observed in the type of waters containing those before pickling and tanning. In these cases survival of Daphnia magna Straus samples constituted 80% and 20% respectively which allows to relate these type of waters to β-M-toxic class, that is moderately contaminated.

Table 1. Characteristics of sewage waters forming under treatment of different types of raw material

Type production process	Characteristics	Sheepskin fur	Furs	Deer paws	Bear skin
Soaking	pH	6,5	7,1	7,0	6,7
	Na_2SiF_6, g/l	0,1	0,4	0,6	-
	COD, mgO_2/l	4268,4	1829,3	1456,9	325,0
	BOD_5, mgO_2/l	1431,6	894,8	480,5	117,0
Degreasing	pH	7,9	7,3	-	6,8
	SSAS, g/l	2,7	1,1	-	4,9
	COD, mgO_2/l	4645,1	2624,3	-	910,7
	BOD_5, mgO_2/l	1830,6	1020,6	-	526,8
Pickling	pH	2,8	2,5	2,3	2,1
	Acidity, g/l	2,5	5,3	10,4	15,9
	Sulphate-ion, mg/l	2,9	-	3,3	2,9
	Chlorides, mg/l	38771	36044	36123	39196
	COD, mgO_2/l	2642,4	1016,3	2638,2	1626,1
	BOD_5, mgO_2/l	816,2	240	644,8	308,8
Tanning	pH	3,5	3,5	2,8	2,5
	Chrome (VI), mg/l	30,0	2,5	2,3	16,0
	Chrome common, mg/l	600,0	320,0	766,0	680,0
	Chrome (III), mg/l	570,0	317,5	763,7	664

The considerable inhibiting influence was produced on the test-objects in the type of waters after the process of degreasing the Daphnia magna Straus loss being observed during one or two hours, which allows to relate the very type of waters to hypertoxical that is highly contaminated. High level of toxic contamination was also characteristic of the waters after soaking. The loss of test-cultures in this case is evidently connected with the presence of siliconfluore of sodium in the water used in the soaking bath for inhibiting of growth and development of putrefactive microorganisms.

According to our concepts one of the lines leading to the solution of this problem lies in the direction of introducing biotechnological method of sheepskin fur treatment based on the combination of the conventional methods of treatment with microbiological ones.

At present different ways of technological processing of sheepskin fur materials have been developed on the basis of using ferments of various nature. A well-known way of degreasing of sheepskin fur is carried out with application of ferment preparation

Protosubtilin G3X for removal of organical substances. Sheepskin materials are usually degreased in the solution containing anionactive synthetic active substances in proportion 1:10 in the total concentration 8-12 g/l, carbonate sodium – 0,5 g/l, at the temperature of 42°C, during one hour. In the case of this degreasing composition being used the quality reactions to the overgreasing of hair cover sheepskin are negative. However on degreasing with ferments of organic origin a slight loosening of hair tension with skin tissue takes place which is not desirable for fur skins [6].

The technology we proposed is based on the use of cultural liquid containing ekzoferment produced by microorganisms into the environment. We used the culture *Pseudomonas sp.* isolated from sewage water after the emulsion degreasing. The identification of microorganism was conducted with conventional methods on the basis of studies of the combined morfologo-cultural and physiobiochemical characteristics [7, 8].

The culture under investigation is a short, gram-negative, separate or disposed in short chainlike active bacillus. On the meatpepton agar colonies were of large size (d=4-5 mm), milk-white colour, smooth brilliant surface, flat-convex half transparent and of slisy consistence. On the meatpepton broth on incubation in thermostat (37±0,5°C) during 24 hours they grew quickly forming thin knobby film of milkwhite colour and a slight flake like sediment.

Pure cultures were kept up on the synthetic agarised nutrient medium including (g/l): Na_2HPO_4 – 1,0; NH_4NO_3 – 1,0; KCl – 0,5; $MgCl_2$ – 0,1. To obtain dense medium agar with 1,5% proportion to the total volume of liquid. SSAS of different chemical nature – 0,5 g/l and organic fat – 2 g/l served as the source of carbon. The following substances non-iogenic – Prevocell W-OF-7, anionactive – Sulphonol NP-3 were used as synthetic surface active substances. The active reaction of medium was kept up during the whole period of cultivation in the limits 7,0. Cultivation was carried out during 5 days in the 250 sm^3 Erlenmeior flask with 150 sm^3 medium on "Shaker-Type 357" (of the round swinging type) at the temperature (37±0,5°C). Increase of biomass was estimated by optical density of cultivated liquid with 540nm on KFK-2.

To define the activity of ekzoferments being produced by cultures the cultivated liquid was used after the precipitation of cells by centrifuge with 5000 rev/m during 10 minutes. To isolate products of organic activity of microorganisms the process of salting was introduced through cultivated liquid of ammonium sulphate. After that the sediment was separated from fugate and in the fugate portion concentration of SSAS was measured. The sediment formed after salting was dissolved in the distilled water and both proteolytic and lipolytic activities were measured. The determination of the quantity of the produced protein in the cultivated liquid has been conducted through biuret method according to Yarosh [9]. Lypolitic activity of eczoferments was estimated according to the modified method of Ott, Yamada [9]. As a lipase activity unit was taken such number of ferments which discharges 1 mk mol of olein acid from 40% of olive oil at pH 7,0 and temperature (37±0,5) °C during one hour. Proteolytic activity has been estimated by the method described by Vilshtetter and Valdshmidt-Leitz. As a proteolitic activity unit was taken the quantity of ferment which forms 1 mg of nitrogen per hour in the suggested conditions of the test [9].

It is well known that optimal removal of organically washed away substances from the hair surface and skin tissues of sheepskin fur is closely connected with the formation of foam, reduction of surface tension and emulgation of hydrotobic substrates. This is achieved in the solution containing SSAS with the concentration of 6-8 g/l and sodium carbonate which allows to create low alkali medium. However the increase of ion-hydroxile results in the loosening hair tension that may become the reason of such defect as hair wearing off the surface.

A number of authors [10] suggest to use lipase which destruct organic admixtures for the removal of greases from the hair covered surface and skin tissues of fur sheepskin. However the use of pure lipase prevents the complete removal of fat substances off sheepskin tissues which can be explained by protein cover around them. That is why for the solution the problem it is necessary to apply a complex of ferments having both lipolitic and proteolitic activity. Herber M.I. stresses that for the maximum accumulation of ferments in the culture optional conditions should be strictly followed that provide optional growth and progress of microorganisms: the full value of nutrition medium, appropriate amount of oxygen, optional pH of medium and temperature [11]. The results of the research are presented in table 2.

Table 2. The influence of SSAS nature on bacterial suspensions

Medium parameters	The duration of cultivation, an hour						
	0	12	24	48	72	96	120
	Prevocell W-OF-7						
pH	6,9	6,9	6,9	7,1	6,9	7,0	7,0
Optical density, D_{540}	0,32	0,5	0,71	1,6	0,92	0,9	0,75
Concentration of SSAS, g/l	0,48	0,32	0,18	0,14	0,08	0,06	0,04
Concentration of protein, g/l	-	2,7	2,8	3,3	3,05	3,05	2,8
Lipolitic activity, unit/g	-	8,15	11,42	18,48	21,63	25,14	22,86
Proteolitic activity, unit/g	-	1,1	2,1	3,6	4,4	4,2	3,2
	Sulphonol NP-3						
pH	7,0	6,9	7,0	7,0	6,8	6,8	6,9
Optical density, D_{540}	1,5	1,7	4,2	6,0	4,4	2,4	2,0
Concentration of SSAS, g/l	0,47	0,44	0,41	0,37	0,32	0,29	0,25
Concentration of protein, g/l	-	1,6	2,6	3,55	3,8	4,8	4,5
Lipolitic activity, unit/g	-	10,63	12,69	14,28	15,78	20,21	22,23
Proteolitic activity, unit/g	-	1,08	2,3	2,5	3,1	3,1	3,3

In the above presented table the results of research illustrate the comparison of the capacity for degrading of two types of SSAS of different chemical property: anionactive – Sulphonol NP-3 and non-iogenic – Prevocell W-OF-7. It was established that the utilization degree of these two types of SSAS considerably differed. Thus the destruction degree of non-iogenic SSAS during 120 hours constituted 92% whereas during the same period of time the destruction degree of Sulphonol NP-3 was 47%. It is probably connected with the fact, that the Prevocell W-OF-7 molecule had been subjected to degrading on two sides – oxyethulchain and hydrocarbon radical. In this case maximum utilization of SSAS was reached almost during 72 hours of cultivation which was proved by considerable growth of dimness in the cultivated liquid during the time period of 48-72 hours.

Analyzing dynamics of growth and progress of the culture under study it can be observed that there is a certain similarity in the consumption of SSAS of both non-iogenic and anionactive chemical nature. The adaptation period for two variants constituted 12 hours. As microorganism became adapted to the medium they began propagating with the growing speed and there appeared evidence of great number of products of organic activity which was proved by the increase of protein concentration in the medium. Intensive utilization of SSAS during 72 hours leads to the limitation of substrate that is exhaustion of nutrition medium and accumulation of the products of metabolism resulting in the phase of autolysis which shows evidence of the reduction of dimness of the cultivated liquid.

It is well known that SSAS can manifest as effective agents influencing the penetration of membranes and cells in general. This means that in the result of the treatment of cells by detergent the ferments become totally accessible for the effect of substrate [12]. As a result of the treatment of SSAS cells lipase which is connected with cell structures is likely to pass into the solution thus explaining the growth of lipolitic activity of the produced ferments during the whole period of cultivation. In this case anionactive SSAS considerably influence the citoplasmatic membrane destructing its integrity. Slightly lower concentration of protein and its lipolitic activity in the cultivated liquid in the presence of non-iogenic of micelle originated by NSAS in water environment their size being greater than of those originated by ionogenic SSAS. Thus the conducted research of fur sheepskin of biotechnological degreasing of fur sheepskin revealed that for the synthetic of a ferment complex with lipolitic activity prevailing it is necessary to apply aerating, pH approximately 7,0, the presence of SSAS in the medium and wool fat as well. On following these above–mentioned conditions the best characteristics can be obtained for the resulting bacterial suspension during the process of growing of the stamm-produced on the developed medium under investigation. Synthetic media for cultivating of microorganisms were prepared in accordance with commonly accepted microbiological methods [13]. The compositions of degreasing solutions are illustrated in table 3.

Table 3. Composition of degreasing solutions

Composition of media	Method of degreasing						
	Typical method	Testing methods					
		Prevocell W-OF-7			Sulphonol NP-3		
		1:0	1:1	1:4	1:0	1:1	1:4
Variants of degreasing solutions	1	2	3	4	5	6	7
Concentration of the complex product of organic activity of microorganisms, g/l	-	5,5	2,3	1,25	4,5	2,5	1,1
Concentration of SSAS, g/l	8,0	0,5	0,27	0,13	0,48	0,26	0,12
Formaldehyde, ml/l	0,5	-	-	-	-	-	-
Carbonate of sodium, g/l	0,5	-	-	-	-	-	-
CMN, cl/ml	-	$5{,}3\times10^{7}$	$3{,}6\times10^{5}$	$2{,}1\times10^{3}$	$2{,}3\times10^{7}$	$1{,}8\times10^{4}$	10^{2}

For the process of degreasing samples of fur sheepskin were selected after soaking performed according to the conventional methods, method of asymmetrical flesh. Previously prepared synthetic media contained both mineral and organic components. Into the media Prevocell W-OF-7 or Sulphonol NP-3 were introduced as SSAS. The process of biotechnological degreasing was performed on the basis of application of concentrated bacterial suspension on *Pseudomonas sp.* basis and the given solution previously soluted two or four times. As a control variant the process of degreasing was performed in accordance of the Unified methodology of fur sheepskin treatment [14]. The content of fat substances in the hair cover and skin tissues were determined prior and after the process of degreasing [15]. The process of degreasing was performed according to the following parameters: duration – 1 hour; liquid coefficient (LC) 10; temperature of test bath (37±0,5)°C; pH about 7,0. The results of investigation are presented in the table 4.

Table 4. Content of greasing substances in the fur sheepskin prior and after the process of degreasing

Degreasing compositions	Content of greasing substances, %	
	In the hair cover	In skin tissues
Prior the treatment	13,25	21,37
1	2,93	6,89
2	3,01	6,9
3	3,1	13,87
4	3,8	14,96
5	2,83	12,41
6	3,11	12,55
7	3,26	14,6

The analysis of the represented data shows that practically all the solutions possessed degreasing characteristics which has been proved by the reduction of organically washed off substances from 13,25% up to 2,8-3,8% and from 21,37% up to 6,9-14,96% both in the hair cover of sheepskin and skin tissues respectively. Greasing substances were being removed in the case of the concentrated and diluted suspension (1:2). In the case of these variants optional removal of grease substances from hair in the limit of 2,0% was observed. In all variants optimal removal of organically washed off substances from skin tissues was gained which is likely to be connected with the presence of ferments with proteolitic characteristic in the solution allowing to destruct protein envelope around greases in the skin tissue.

According to the organoleptic estimation the hair cover of fur sheepskin samples looked friable, white, degreased and the hair loosening was not observed.

Thus the performed research resulted in the development of bacterial suspension on the basis of *Pseudomonas sp.* culture and SSAS of different chemical nature with considerable reduction of SSAS consumption in the degreasing compositions from 8 up ton 0,25 g/l, complete elimination of formaldehyde and sodium carbonate which will allow to reduce considerably the level of toxic contamination of sewage waters with preserving of quality characteristics of finished articles.

REFERENCES

[1] Rodionov A.I., Klushin I.N., Goroshechnikov N.S. Technique and the protection of the environment. – M.: *Chemistry*, 1989. - 512 p.

[2] Stadniskiy G.V., Rodionov A.I. Ecology: Textbook for chemicotechnological institutions. – M.: *Higher School Publishers*, 1988. – 272 p.

[3] The combined application of ecobiotechnological consortium of microorganisms and easily oxydised SSAS in the technological processes /Shlbouyev D.V., Dumnov V.S., Ineshina E.G., Tzhirenov V.Zh // Volume of articles of the First International Conference "Industry. Technology. Ecology" (ITE'98). – M.: "*Stankin*", 1998.- P.188-191.

[4] Methods of bioindication and biotesting of native waters. Ed.2.- L.: *Gidrometeostat,* 1988.- 275 p.

[5] Braginskiy L.P. On principles of classification of freshwater ecosystems //*Gidrobiologicheskiy magazine*, 1985, V.21, #6, P.65-74.

[6] Kurisheva G.N., Zueva V.G. Application of proteolitic ferment preparation on treatment of fur// Vol. of scientific works "*Research in the field of fur, increase of quality of fur material and half-finished products*".- M.: 1987, P. 8-21.

[7] Methods of general bacteriology / Ed. by Gerhard F and others.- M.: *Mir*, 1984, V. 3, -264 P.

[8] Determinater of bacteria by Berdgie / Hoult G., Kring N., Snith P and others. – M.: *Mir*, 1997. – 1 and 2 vol.

[9] Laboratory manual and workbook on technology of ferment preparation /Grachev I.M., Grachev U.P., Mosichev M.C. and others.- M.: *Legkayay I Pichevayay Promishlenost*, 1982.- 240 P.

[10] Investigation of ferment process of degreasing fur sheepskin /Fomina L.A., Rohvarger O.D., Zubin A.M. and others.- *Vol. of scientific works VNIIMP* – M.: 1980, P. 3-7.

[11] Gerber N.I. Production and application of ferment preparations.- Kiev: *Naukova Dumka*, 1978.- 103 p.

[12] Eliseev S.A., Kucher R.V. Surface active substances and biotechnology.- Kiev: *Naukova Dumka*, 1991.- 116 P.

[13] Netsepliaev S.V., Pankratov A. Ya. Laboratory manual and workbook of food products of animal origin.- M.: *Agropromizdat,* 1990.- 223 p.

[14] Basic technology of fur sheepskin. – M.: *CNIITEI,* 1988.- 153 P.

[15] Golovteeva A.A. and others. Laboratory manual on chemistry and technology of leather and fur /Golovteeva A.A., Kutsidi D.A., Sankin L.B. – M.: *Legkaya I Pishevaya Promishlenost,* 1987. – 312 P.

In: Industrial Application of Biotechnology
Editors: I. A. Krylov and G. E. Zaikov, pp. 105-113
ISBN 1-60021-039-2

Chapter 13

INCREASING EFFICIENCY OF BIODEGRADATION OF FAT-CONTAINING WASTES OF MEAT-PROCESSING INDUSTRY

N. A. Suyasov*, B. A. Karetkin, S. V. Kalyonov, I. V. Shakir and V. I. Panphilov
Mendeleyev University of Chemical Technology of Russia, Moscow, Russia

ABSTRACT

The work is devoted to the elaboration of the technology of biodegradation of fat-containing wastes of meat-processing industry. The researches carried out have shown that the most perspective biodestructor of such substrates is microorganism *Yarrowia lipolytica*. It is established that the preliminary yeast adaptation to metabolites of natural micro flora, the optimum age and concentration of sowing material, carrying out the process at sufficient aeration, pH 5.0-5.5, and the temperature about 30 ^{0}C allow to reduce the duration of cultivation less than 48 hours and to increase the crude protein contents up to 55-60 %.

Keywords: fat-containing wastes, meat-processing industry, Yarrowia lipolytica.

INTRODUCTION

Processing industrial wastes is one of the major problems, a great deal of attention being paid to its decision recently. It is known a large number of ways of converting mankind economic activity wastes based on mechanical, thermal and chemical effects. Besides biological conversion has a wide expansion as it frequently allows not only to decrease harm considerably to the environment but also to get some products.

* Mendeleyev University of Chemical Technology of Russia, 9, Miusskaya sq., Moscow, 125047, Russia e-mail: nik-suyasov@mail.ru

Annual accumulation of a huge quantity of wastes of meat-processing industry enables to recommend them as cheap substrata for microorganism cultivation processes [1]. The firm phase collected in the grease skimming tank of sewage disposal unites of meat-processing plants is determined to consist of fat (40-45 %), protein (30-32 %) and other components. Substantial amounts of the wastes quickly rotting with formation of unpleasant smells are inevitably formed during work of meat-processing plants. The presence of fats in structure of wastes results in formation of dense sedimentation on the walls of pipes and in tanks [2-4]. According to up-to-date requirements industrial general water wastes are to be preliminary cleared on local sewage disposal unites before discharge in sewage systems to remove the pollutions interfering transportation and to purify them biologically. There is an inevitable accumulation of fat-containing wastes at meat-processing plants and therefore a problem of their recycling arises.

Now a number of technologies allowing to utilize or process fatty wastes is offered, among them it is possible to allocate two basic directions: physico-chemical and microbiological ones. The first one includes: dividing into fractions [5], realization of alkaline hydrolysis and oxidation [6], crushing and thermal processing with steam with further use as fodder additives [7], ozonization of the sewage with subsequent coagulation and flotation [8] etc. Among microbiological methods of destruction of fat-containing wastes it is possible to allocate the use of biological objects, which include live cultures of microorganisms, and also a complex of enzymes, with subsequent oxidation by potassium permanganate and hydrogen peroxide [9]. Also carrying out biodestruction by anaerobic degradation and the usage of biofilters are offered [10]. One of the most perspective ways of biodegradation of wastes of meat-processing industry is an aerobic cultivation of high protein containing cultures of microorganisms because the biomass formed can be used as fodder additives for increasing nutriency of fodder diets for agricultural animals. Vegetative forages do not always contain necessary quantity of proteins, essential amino acids, vitamins, therefore their enrichment by additives is necessary [1].

Materials and Methods

Cultivation was carried out in flasks of volume of 250 ml (with volume of nutrient medium- 100 ml) with constant stirring with intensity of 150 rpm and at the temperature 30 ^{0}C, and also in a bioreactor (volume - 5 l) filled with nutrient medium till 70 % at the temperature 20-37 ^{0}C and constant hashing with intensity of 250 rpm, with pH being supported at a level of 5.0 – 5.5 by the use of water solution of ammonia.

Also cultivation was carried out on a solid nutrient medium containing some mash, meat extract, pork fat, wastes of meat-processing industry as a source of carbon. During growth the lipolytical activity of cultures of microorganisms was analyzed as quantity of micromoles of olein acid released during olive oil emulsion hydrolysis by 1 mg of biomass during 1 hour [11].

Growth of microorganisms was estimated by direct calculation with both Gorjaev's chamber and gravimetric analysis.

During researches the contents of crude protein (Kyeldal method), true protein (Barnshtein method), general fat (Sokslet method), carbohydrates (Dubua method), and nucleonic acids (Spirin method) of biomass obtained were analyzed [5, 12].

RESULTS AND DISCUSSION

The ability of a number of microorganisms to produce lipases lays in the basis of a microbiological way of biodegradation. As far as nowadays there are already examples of using micro flora, for example, for separate types of wastes at the realization of microbic remediation of industrial sewages [13], we have carried out the research on their isolation.

As a result two autochthonous cultures for utilization of fat-containing wastes were isolated (Table 1), one of which being referred to bacteria. This bacterium was established to be Gram-positive one. The research of growth characteristics of the culture given has shown a low efficiency of its use in biodestruction of fats.

Table 1. Characteristics of the native microflora

Characteristics	*Yeast culture*	*Bacterial culture*
The form of a colony	Round	Round
Edge of a colony	Wrong	Smooth
Structure of a colony	Fibrous	Homogeneous
Surface of a colony	Rough	Round
Shine and transparency of a colony	Matte	Brilliant
Consistence of a colony	Soft, it is easily removed from an agar	Mucous, it is easily removed from an agar
Colour of a colony	White	Colorless
The form of a cell	Rectangular with rounded corners	Sticks
The size of a cell, micron	13,75 x 5,5	1,4 x 0,5
Relative positioning of cells	Separate cells, chains, micelium	Separate cells
Gram-type	-	Gram-positive
Ability to form disputes	Hlamido-disputes and ballisto-disputes are not revealed	Endo-disputes are found out
Sensitivity to antibiotics:	Low-sensitive to the majority of tested antibiotics	High-sensitive to the majority of tested antibiotics
Endoplasmatic inclusions	Granulose, starch, volutine are found out	-
Fermentive activity	Capable to destruct raffinose, maltose, saccharose, lactose, galactose, glucose.	-

The second microorganism was suggested to be related to the yeast as it is capable to destroy sugar, to form the similarity of pseudomicelium, and to exist as separate large cells.

The cultures allocated were tested on ability of assimilation both a standard source of carbon (glucose) and specific substrata (pork fat and wastes of meat-processing industry).

Thus, it has been revealed that the physiological optimum of the growth for both cultures is the temperature about 30^0C.

At a more detailed studying of the yeast culture it was established that during the growth the cells change their form essentially. Originally they form hyphae which are transformed a bit later into pseudomicelium that is a characteristic for the yeast, and then it divides into separate cells. Besides, the cells that divide by budding and binary bisection were found out.

The research of characteristics of the yeast culture growth was carried out in a laboratory bioreactor, the firm phase of the wastes sustained in the grease skimming tank within 1, 5, and 13 days were used as a source of carbon. As a result, the maximum yield of the yeast was established to be reached at cultivation with 1 day wastes and their concentration (13.54 g/l by the 42-nd hour) is 1.5 and 2 times more than the concentrations at the use of 5 day and 13 day wastes, accordingly. During fermentation a lipolytical activity was shown by the culture discretely, the maximal its values being reached at the time of termination of *log*-phase and in the middle of exponential phase (by the moment of disintegration of hyphae in separate cells). The analysis of the biomass structure has shown a preferability of fresh wastes as they allow to achieve 45 % of true protein contents, and the amount of fatty substances does not exceed 12,5 % (Table 2). It allows relating the product to a high protein containing one.

A possibility of application of microorganisms - producers of lipases in processes of biodegradation of fats has been also investigated. Representatives of basic systematic groups have been tested: bacterial strains - *Bacillus mesentericus*, *Bacillus subtilis*, *Acinetobacter sp.*; fungus - *Aspergillus orysae*, *Penicillium orysae*, and also yeasts - *Candida scotti*, *Yarrowia lipolytica*.

Table 2. Characteristics of biomass of the native yeast and bacteria obtained at cultivation with various substrata

Characteristics of a biomass	Substratum			
	wastes of the 1-st day *	wastes of the 5-th day *	wastes of the 13-th day *	wastes of 1-st day **
The maximal accumulation of biomass, g/l	13,5	9,1	6,4	2,5
Colour of biomass	light brown	light brown	grey	grey
True protein, %	44,1-45,3	28,0-29,2	19,3-20,1	19,4-21,2
Crude protein, %	53,2-55,1	34,4-35,1	23,2-25,1	23,0 - 25,3
Fat substances, %	11,0-12,5	15,3-19,7	25,0-26,5	45,0-48,1
Carbohydrates, %	20,0-23,0	17,1-14,0	13,0-14,5	24,0-25,5
Nucleic acids, %	4,0-5,1	3,0-4,5	1,3-2,5	0,4-0,7
The test for sharp toxicity	not toxic	not toxic	not toxic	not toxic

* - Yeast culture

** - Bacterial culture

The research of characteristics of the chosen cultures of microorganisms was carried out on nutrient medium in presence of both glucose as a source of carbon and soya and fusible fraction of pork fat as stimulators of lipolytical activity. As a result it has been established that *Yarrowia lipolytica* essentially surpasses all the other cultures in lipolytical activity and in yeild at cultivation on pork fat, conceding only *Candida scotti* in the contents of cellular proteins. Preferability of use of yeast *Yarrowia lipolytica* also coordinates to the literary data that the culture concerned is capable of high growth on fat-containing nutrient medium [14].

At studying features of the lipolytical fermental complex of yeast *Y. lipolytica* it was established that the given culture contains different lipases, distinguished by both the optimum values of acidity of the medium (acid and neutral lipases) and substratum specificity.

However, the research of growth of the microorganisms concerned on wastes of meat-processing industry has shown that the process given has such shortages as a high duration of cultivation - up to 120 hours (Figure 1) and poor quality of the biomass obtained (protein contents does not exceed 30,7 %).

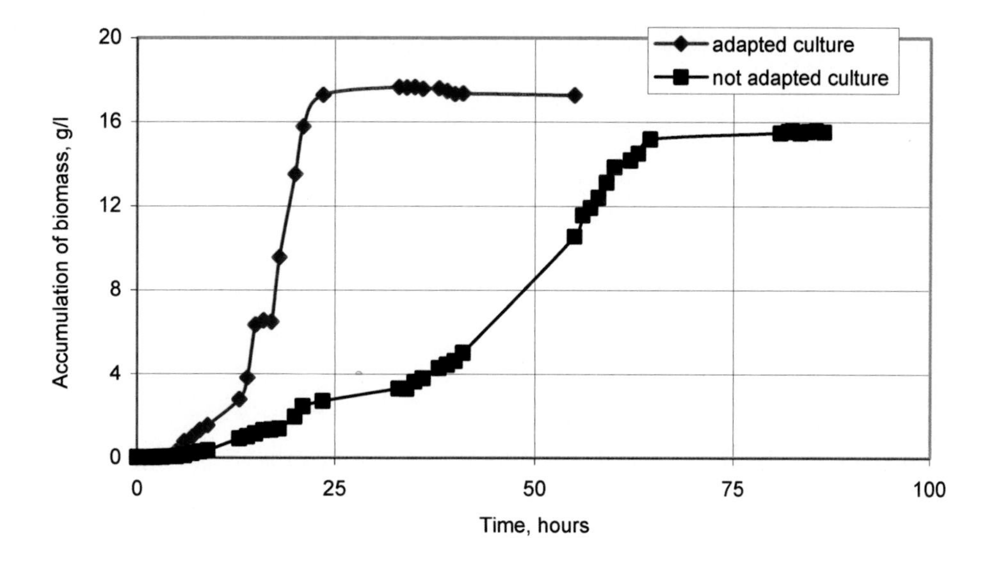

Figure 1. Accumulation of biomass during fermentation *Y. lipolytica* on fresh fatty wastes (20 g/l)

One of necessary stages of our research was to study samples of the wastes which were sampled during several cycles of functioning of the grease skimming tank of meat-processing plant, thus the time of their presence in the grease skimming tank varied from 1day till 13 days. It was revealed that during storage the mass content of crude fat is essentially reduced (from 98.4 % to 28.3 %), and also the contents of proteins is increased (Table 3) that testifies the presence of natural microflora in the grease skimming tank. The change of the acid number allows assuming that the indigenous microflora consumes readily available fats with low molecular weight and thus increases own weight, displacing the ratio of fat and other organic derivatives.

It is known that for sowing it is expedient to use a sowing material being at exponential phase of a growth; however at realization of process of biodegradation of fats it is necessary to be guided also by lipolytical activity of the culture. The analysis of the literature has shown that *Yarrowia lipolytica* shows the maximum lipolytical activity at the stationary phase, however the researches carried out earlier testify that lipase of given strain is also active at adaptation phase. During investigation of the influence of sowing material pretreatment on efficiency of biodegradation of fatty wastes it was established, that it is optimum to make the sowing with the culture which is at the beginning of the exponential growth phase, at the initial contents of cells in a nutrient medium not lower than 0.2 million/ml, i.e. the amount of sowing material does not exceed 1 % of volume of nutrient medium in the bioreactor. If the wastes stored in the grease skimming tank from 5 till 13 days (of the 5-th day wastes and the 13-th day wastes) are used as a substratum then the amount of sowing material should be a little bit higher.

Table 3. Comparative characteristics of structure of fat-containing wastes in dependence on the time of storage in the grease skimming tank

Analyzed sample of wastes	Proportion of crude fat, % by weight	Proportion of readily available fat, % by weight	Acid number of fats	Crude protein, %	Carbo-hydrates %	Nucleonic acids, %
1 day of storage	96,9-98,4	92,2-94,3	190,1-219,4	12,5-13,9	1,2	0,4
5 days of storage	70,3-76,6	63,3-74,6	131,4-140,3	17,5-21,0	11,1	0,3
13 days of storage	28,3-31,4	26,1-28,3	118,8-124,6	19,0-42,1	22,6	0,3

The research of the influence of waste storage on the yield of cultures has shown that concentration of 20 g/l is optimum for a more full consumption of the substratum, provided that the wastes of meat-processing industry are fresh. At cultivation on 5-th day wastes and 13-th day wastes the optimum concentrations are of 14 g/l and 7 g/l, accordingly, and the further increasing of substratum concentration results in decreasing the growth of the culture. Thereupon, the assumption was made that the native microflora during growth does not only change the structure of substrata, but also accumulates products which oppress the growth of yeast *Yarrowia lipolytica*.

With the purpose of confirming the assumption involved the research of native microflora metabolite influence on the growth of *Yarrowia lipolytica* was carried out. Thus, the metabolites of the bacterial native culture were established to be able to slow down the yeast native culture growth and to reduce the yield of the biomass.

One of the ways allowing decreasing the influence of metabolites of native microflora is carrying out an adaptation of the microorganisms selected. It is established that the yield of biomass is increased up to 52 % by the 7-th passage, the duration of cultivation being reduced from 120 hours to 48 hours. The analysis of the biomass has shown that the contents of protein in biomass grow from 24 % up to 45 % by dry weight. But at storage on solid nutrient mediums and at the subsequent deep cultivation of the adapted culture the growth characteristics are not reproduced completely, the decrease of the growth rate being observed. However, solid nutrient mediums contained wastes especially the 13-th day sustained in the grease skimming tank rendered a stimulating influence on productivity of the culture.

The research of morphological features of the adapted strains *Yarrowia lipolytica* has shown that the cells are more homogeneous by the form, its size changing a little, and microorganisms became steady against the action of some antibiotics.

For studying of complex influence of the chosen conditions the process of biodegradation of fat-containing wastes was carried out in a laboratory bioreactor, the duration of the *log*-phase being reduced from 18 hours to 4 hours, the specific rate of growth increasing in four times. The yield of culture increased till 12 % and it reached 17.5 g/l already by the 25-th hour (Figure 1).

Besides it was established that the specific lipolytical activity of *Yarrowia lipolytica* adapted to the wastes of meat-processing industry reached 960-742 unites/mg in the interval from the 4-th hour till the 11-th hour, i. e. almost three times higher than that at the similar conditions of cultivation but with the use of the sowing material that is not adapted. Thus, it is possible to make a conclusion that the realization of adaptation of culture not only raises stability of microorganisms to metabolites of native microflora, but also it essentially increases the lipolytical activity and also influences on the characteristics of growth.

It is necessary to notice that the diauxie is observed in process of biodegradation of animal fats by *Yarrowia lipolytica* for the adapted culture from the 16-th hour till the 19-th hour, and for not adapted one from the 33-th hour till 40-th hour.

For revealing a sequence of consumption of the basic components of lipids, the growth of *Yarrowia lipolytica* on various substrata in which were used glycerin, stearin acid, and olein acid as the only source of carbon was investigated. Thus, it was established that the longest phase of adaptation of the culture was observed with olein acid, and the minimal *log*-phase - at cultivation on the nutrient medium containing glycerin. It is possible to assume that first of all there is a consumption of glycerin, then stearin acid and the last one – olein acid at assimilation of animal fats.

Also we investigated a possibility of increasing the yield of the culture by the carrying out a periodic cultivation with additional charging of substratum. Thus, it is possible to achieve a concentration of a biomass of 23.8 g/l by the 62-nd hour that surpasses the yield of cultivation without additional charging up to 25 %. The proportion of the true protein makes up about 40 % that is much higher than that one at the use of not adapted culture. However, the best growth and quality of biomass were observed at standard periodic cultivation with the use of adapted *Y. lipolytica* at which the growth rate makes up 0.32 $hour^{-1}$, the proportion of the true protein is about 50 %, and the concentration of fats in a biomass does not exceed 7 % by weight (Table 4).

It is necessary to note that biodegradation of fatty compounds is dealt with the heterophase cultivation in which consumption of a substratum is carried out on boundary of phases. One of ways of increasing an efficiency of consumption of nutrients is to increase the area of the substratum surface. It can be reached due to increasing the dispersiveness of the system by processing by ultrasound. In addition ultrasonic waves are capable to stimulate processes of oxidation of compounds that allows carrying out preliminary partial degradation of the substratum and raises its assimilability by the microorganisms. Besides ultrasonic processing of fatty wastes is capable to replace technical sterilization and decrease essentially the contents of viable microorganisms of natural microflora.

During the work an ultrasonic device with frequency of 25 kHz was used. Conditions of processing have been chosen by researching the fatty dispersion formed in the mineral solution. It is necessary to note that the essential increase of optical density (at 540 nm of wave-length) of dispersion with following going out to a plateau was marked with the increase of duration of ultrasonic processing. The initial temperature rising in the ultrasonic bath makes possible to get maximal values of optical density for less time. Thus, the optimum conditions of processing by ultrasound were chosen: the initial temperature of water in the ultrasonic bath - 32 – 34 ^{0}C, the duration of processing - not less than 7 minutes (Table 5).

These specified features of fat-water dispersion were observed at concentration of pork fat in an interval from 1 to 50 g/l.

Table 4. Characteristics of growth and structure of the biomass of yeast *Y. lipolytica* obtained at cultivation on fat-containing wastes

Parameter	**A**	**B**	**C**
log-phase, hours	18	4	12
Specific growth rate, h^{-1}	0,08	0,32	0,15
Maximal accumulation of a biomass, g/l	15,5	17,5	23,8
Output, g/g	0,78	0,88	0,63
Colour	light yellow	light yellow	light yellow
Smell	peculiar to yeast	peculiar to yeast	peculiar to yeast
Crude protein, by weight, %	31,5-33,5	57,8-60,3	44,5-49,8
True protein, by weight, %	26,5-28,2	48,6-50,7	37,7-41,8
Carbohydrates, by weight, %	11,1	12,7	17,3
Crude fat, by weight, %	11,6	6,7	13,5
Nucleonic acids, by weight, %	2,5	3,5	2,6
Result of the test for sharp toxicity	not oxical	not toxical	not toxical

A - periodic cultivation of *Y. lipolytica*;
B - periodic cultivation of adapted *Y. lipolytica*;
C - periodic cultivation of adapted *Y. lipolytica* with additional charging of substratum.

Table 5. Accumulation of the biomass of the yeast at cultivation on the fat-containing substratum

Characteristics of fermentation					
Initial concentration of substratum, g/l	Pretreatment of the substratum		Time of cultivation, h	Accumulation of biomass, g/l	Output, g/g
	Duration, min	Initial temperature, ^0C			
1	Is absent		96	0,96	0,96
1	7	32 - 34	72	1,38	1,38
10	Is absent		96	8,14	0,81
10	7	32 - 34	72	12,56	1,26
30	Is absent		96	19,28	0,64
30	7	32 - 34	72	28,50	0,95
50	Is absent		96	22,71	0,45
50	7	32 - 34	72	32,23	0,64
10	6	32 - 34	72	9,85	0,99
10	8	32 - 34	72	12,57	1,26
10	7	23 - 25	72	10,05	1,01
10	7	44 - 46	72	12,55	1,26

The use of ultrasonic pretreatment allows increasing bioavailability of a fatty substratum essentially due to the essential increase of the area of the surface. Besides, ultrasonic pretreatment of a nutritious medium is capable to replace technical sterilization and to oxidize fats partly.

The results received testify that the realization of ultrasonic pretreatment of a fat-containing nutrient medium allows to increase accumulation of biomass essentially and to reduce the duration of cultivation up to 72 hours at the initial temperature of processing 23 – 44 ^{0}C, within not less than 5 minutes, and the best result is observed at the initial temperature of pretreatment not less than 30 ^{0}C, within not less than 7 minutes.

Thus, the researches carried out have shown that aerobic biodegradation of fat-containing wastes of meat-processing industry is effective and the yeast biomass obtained can be considered as high protein containing one.

REFERENCES

[1] V.A.Krohina. Mixed fodders, fodder additives for agriculture animals. *Agropromizdat*, Moscow, 1990. 304 p. (in Russian).

[2] A.J.Martynov, L.L.Nikiforov, G.S.Rudenko: *The meat industry*, 8, 21 (2003). (in Russian).

[3] A.N.Ivankin, R.V.Iljuhina: *The meat industry*, 5, 46 (2001). (in Russian).

[4] V.M.Kovbasenko: Wastes of meat-packing plants. *Pishchepromizdat,* Moscow, 1989, 237 p. (in Russian).

[5] A.N.Ivankin: Creation of the system of biotransformation of seldom used fatty wastes in production of increased biological value. *MGUL*, Moscow, 2002. 41 p. (in Russian).

[6] Patent of the Russian Federation №2207327 C2, 2001. (in Russian).

[7] Patent of the Russian Federation №94033855 A1, 1994. (in Russian).

[8] Patent of the Russian Federation №97118861 A, 1999. (in Russian).

[9] Patent of the Russian Federation №2001132315 A, 2001. (in Russian).

[10] Patent of the Russian Federation №2161892 C1, 2001. (in Russian).

[11] Patent of the Russian Federation №2148645 C1, 2000. (in Russian).

[12] Yeast fodder. *Test methods:* GOST 28178 - 89. (in Russian).

[13] V.V.Safronov: *Intensification of system of biodestruction of pollutions of low waste concentrated sewages*, Dis. ... dr. 2004. (in Russian).

[14] Tsapina A.V., Gradova N.B., Gornova I.B.: *Scientific researches of the higher school in the field of chemistry and chemical products*, 179, 206 (2001) (in Russian).

In: Industrial Application of Biotechnology
Editors: I. A. Krylov and G. E. Zaikov, pp. 115-121
ISBN 1-60021-039-2

Chapter 14

GAS VORTEX BIOREACTOR. SUMMARY

U. A. Ramazanov**, *V. I. Kislih, A. P. Repkov and I. P. Kosjuk
Joint Stock Company "Sajany", Novosibirsk City, Russia

ABSTRACT

Gas-vortex bioreactor has been created. It uses the absolutely new way of mixing (patents of the USA, Japan, Europe). Practically all types of cells and microorganisms are successfully cultivated in it. Full repeatability of laboratory results at industrial application, low power consumption, mixing of viscous fluids, functioning being filled at 15-90% of volume are its important features.
Large-scale production of vaccines in a 300L gas-vortex bioreactor using embryonic cells (CEF) is being launched at the SandP Holding "VIRION".

Key words: Bioreactor, fermenter, cultivation, *embryonic cells, hybridoma, gas-vortex,* biotechnological production, manufacture.

Accelerated development of creation of new drugs using mammalian and other sensitive cells makes it necessary to develop new efficient apparatuses providing optimal conditions of microbiological synthesis - bioreactors of the new generation. The bioreactor is intended for the creation of the most optimal conditions for vital functions of the cultivated cells and microorganisms.

This is the provision of:

1. Good mass exchange by the gas phase i.e. respiration;
2. Feeding i.e. nutrient substances supply;
3. Metabolite withdrawal.

The cells should not be subjected to mechanical, thermal and other stress impacts.

Two methods of stirring are commonly known and used:

* Joint Stock Company "Sajany". Office 403, 8 Nikolaeva Street, Novosibirsk City, 630090, Russia. sajany@bioreactor.ru, sajany@ngs.ru, tel/fax (+7 3833) 332601, 333369, 309231

- The first method is stirring with a revolving mechanical device put into the liquid phase and making the fluid move (like a spoon in a glass of water).
- The second method is stirring by the gas phase blowing through the liquid one (different airlift and bubble devices).

I. The disadvantages of bioreactors with a mechanical stirrer are as follows:

- Highly turbulent and stagnant zones are formed in the process of stirring, and therefore, nutrient substances supply to the cells is performed unevenly,
- The same occurs with metabolite withdrawal,
- Surface mass exchange in the apparatus is insufficient for many cell and microorganism cultures for the same reason,
- Cultivated cells and microorganisms die as a result of cut exertions arising at the ends of the stirrer blades.

In bioreactors with a mechanical stirrer almost 70% of the consumed power is used to overcome the resisting forces of the medium; mechanical power is transformed into thermal one i.e. harmful overheating of the culture fluid takes place. It becomes necessary to withdraw this excess heat, which requires additional expenses. The energy (temperature) is introduced unevenly within the whole volume, which adversely affects the results of biotechnological processes requiring the work in a strictly limited temperature range.

At the ends of the stirrer blades there appear of local overheating zonules, which are also destructive for cells.

II. Airlift bioreactors have good mass exchange by the gas phase, but non-intensive stirring. The disadvantages of airlift bioreactors as follows:

- Due to weak stirring (i.e. nutrient substances supply and metabolite withdrawal) they not always are suitable for cultures with high vital activity.
- Floating air bubbles destroy or injure sensitive cells, for example, embryonic or insect ones at "bursting".

Besides, abundant foaming takes place in bioreactors of this type. This does not allow the whole volume of the apparatus to be used, and the use of a chemical defoamer decreases the quality of the end product and makes the process more expensive. It is impossible to use viscous culture fluids in airlift bioreactors.

Most of bioreactors used worldwide present a combination of these two types of apparatuses with the above disadvantages, which become apparent to a greater or lesser extent depending on the apparatus design.

These disadvantages are associated with the injury of cells and microorganisms at stirring, insufficiency of mass exchange, the presence of turbulent and stagnant zones, high power consumption and low characteristics at working with viscous media. The gas-vortex gradientless bioreactor uses a different method of stirring.

The gas-vortex bioreactor has no stirrer within the liquid phase and, therefore, does not cause the above-mentioned problems.

The bioreactor (Figure 1) presents a thermostated reservoir, in which the gas civility and the culture medium are freely separated by the culture medium surface. The filling volume can change from 10 to 90% of the bioreactor physical volume depending on the technological requirements.

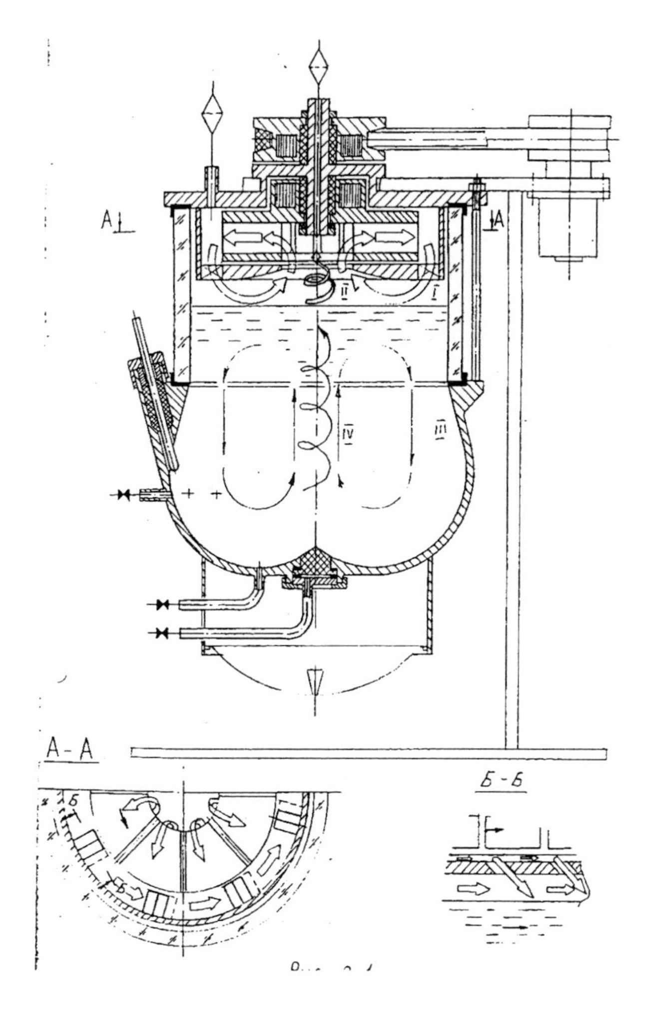

Figure. 1

An intensive air vortex is created in the gas cavity of the bioreactor, over the culture medium surface. Air involves the culture medium in rotary movement, in which the field of velocity with an axial component is formed.

Due to a significant change in the tangential component of velocity in the vortex, there is a difference between statistical pressures at the periphery I (high) and the center II (low). The pressure differential in the gas through the free surface of the culture fluid generates axial motion in the latter: descending in the peripheral zone (III) of the bioreactor and ascending in the paraxial zone (IV).

Stirring of the culture medium in the bioreactor is performed by generating three-dimensional movement of the type of a "rotating vortex ring" (quasi-stationary flow with axial counter-flow) in the liquid medium. The movement is generated by an aerating gas vortex due to the pressure differential over the surface and the force of friction of the airflow against the suspension surface.

The aerating gas vortex is formed above the centrifugal activator installed above the fluid surface and is projected on the liquid phase with a special device.

Three vectors of movement are formed in the fluid:

a) In the horizontal plane,
b) In the vertical plane,
c) There is a radial component.

As a result, smooth but rather efficient stirring without foaming, hydroblows, cavitation, highly turbulent and stagnant zones is achieved.

- The gas-vortex bioreactor has a high rate of mass exchange - 6000-8000 (KL 1/ hour).
- Its characteristics do not change at filling up to 10-90% of the volume, which allows the intermediate "start-up" bioreactors to be removed at industrial production.
- Power inputs for stirring of a fluid with the viscosity of water make up only 0.3 Wt/l, it is 10-12 times less than that consumed by bioreactors with a mechanical stirrer. In the gas-vortex bioreactor, 98% of the input power is used directly for stirring, and the energy (temperature) is evenly introduced within the whole volume. No local overheating zones i.e. zonules of high temperature are formed in the process of stirring.
- The peculiarities of twisted flows provide for the possibility of stirring of highly viscous fluids in the gas-vortex bioreactor (higher than 1.27 P). Polysaccharides with viscosity of 1.27 poises were produced during joint work with the Gubkin Institute of Oil and Gas; this exceeds the viscosity of water by 1270 times.
- Gas vortex is an efficient defoamer;
- The bioreactor hydrodynamics practically does not depend much on the level of fluid in it, and the bioreactor is easily to scale up;
- The above characteristics of the gas-vortex bioreactor allow us:
  - To cultivate cells, poorly reproduced in known types of bioreactors;
  - To start up a bioreactor of the same type but of a larger volume at a ratio of 100:1 between the volumes of the bigger and the smaller bioreactors and to remove bioreactors of an intermediate volume from the technological chain;
  - Not to use a chemical defoamer that makes the further process of purification and production of the end product more complex and expensive;
  - To use the bioreactor in processes involving the use of viscous fluids or the production of such fluids by microbiological synthesis.

Below is the list of data on the cultivation of some cell types:

Table 1.

Cell lines	Cultivation time, h	Cells concentration mln/ml		Number of living cells, (%)
		at the beginning	at the end	
BHK-21	60	0.46	2.9	98%
Mamestra brassicae IZD MB-0503	72	0.4	2.2	96%
Munne myeloma Sp210-Ag $14P_3$	72	1.0	6.0	95%
Human lymphocytes MT-4	96	0.52	2.2	93%

Such cell lines as VERO, A_4C_5 (a hybrid of kidney cell and porcine lymphocyte) and human thymus cells (T-5) were also successfully cultivated in the gas-vortex bioreactor. Twenty- seven million cells were produced within 63 hours of cultivation after loading the norm of seeding of 800,000 chicken embryos fibroblasts into the gas-vortex bioreactor.

A serious problem of the application of many laboratory developments to production is the complexity of scaling up of the obtained results i.e. the reproduction of the laboratory process on an industrial scale. It is often very difficult and sometimes impossible to reproduce it especially with sensitive cells (hybrid, embryonic ones, etc.). A significant portion of promising developments has not been realized so far because it is impossible to reproduce the obtained results at increased volumes of cultivation. This is associated with the fact that the hydrodynamics of the stirring process at scaling up in the existing bioreactors significantly changes depending on increasing the apparatus volume. Problems of this type do not arise at gas-vortex method of stirring.

The use of gas-vortex bioreactors allows us to realize the conception of a highly efficient universal modular biotechnological enterprise i.e. the creation of a biopharmacological complex of a new type.

The essence of the conception of a biopharmacological complex of a new type is that it consists of a continuous highly economical "head" block of gas-vortex gradientless bioreactors allowing cultivation of practically any type of cells and microorganisms and a modifiable modular "tail" block, the modules of which are completed or modified depending on the processed biological product supplied from a block of bioreactors.

Such biotechnological production allows a transfer from one type of producer strain to another one. The enterprise can promptly react to the latest developments in the area of practical biotechnology without large expenses and change or widen the range of the manufactured products.

Cost efficiency of such production is achieved by:

- Minor initial investments as a results of:
  - Reduction of the number of bioreactors in the technological chain;
  - Making communication pipelines simpler and shorter;
- Reduction of total production costs (electric power, steam, detergents, area, wages fund);

- Universality of the facilities, i.e. feasibility of successful operation with different types of producer strains;
- Financial stability of the enterprise as a result of rapid and low-investment implementation of technologies of new biotechnological manufacture;
- Technological effectiveness of production;
- Increased yield of biomass;
- Reduced losses at cultivation.

In our opinion, broad application of gas-vortex bioreactors and the creation of universal biopharmacological complexes of the new type on their basis is one of promising trends in industrial biotechnology.

Sphere of Application

Production of drugs (including ones using highly sensitive embryonic, hybridoma and other cells).

Production of a wide range of microbiological preparations for agriculture and veterinary medicine.

Production of polysaccharides and oil destructors for oil producing industry.

Manufacture of products for food and light industries (ferments, food supplements, etc.).

Production of surface-active substances and ferments for chemical industry.

Industrial apparatuses of gas-vortex type are used to produce a vaccine using embryonic cells.

The photo (Figure 2) shows a gas-vortex bioreactor of 300 L, which is used to produce tick-borne encephalitis vaccine using embryonic cells at SandP Holding "Virion", Tomsk..

The development is protected by patents of RF, USA, Japan and 6 European countries.

Figure. 2

REFERENCES

[1] R. Pohorecki, J. Balduga: New model of micromixing in chemical reactors. *Ind. Eng. Chem. Fundam,* 1983.Vol.22, N 4. – 398-405 p.

[2] U.E. Viestur, N.G. Kristansons, E.S. Bilinkina: Cultivation of microorganisms. *Pishchevaya promishlennost'*, Moscow, 1980. 231 p. (in Russian).

[3] P.P. Loboda, J.V.Karlash: Peculiarities of mass transfer in bioreactors upon intensifying and scaling the processes of microbiological synthesis. *Hydrodynamics and transfer processes in bioreactors/* Edited by R.S. Gorelik. IT, Novosibirsk, 1989. 190 p. (in Russian).

[4] V.K. Shchukin, A.A. Halatov: Heat exchange, mass exchange and hydrodynamics of twisted flows in axially symmetric canals. *Mashinostroenie,* Moscow, 1982. 200 p. (in Russian).

In: Industrial Application of Biotechnology
Editors: I. A. Krylov and G. E. Zaikov, pp. 123-129
ISBN 1-60021-039-2

Chapter 15

NICOTINATE-PHOSPHORIBOSILTRANSFERASE: PROPERTIES AND REGULATION

V. Zh. Tsirenov*, W. G. Dulyaninova, E. M. Podlepa and A. A. Sandanov
East-Siberia State Technological University, Ulan-Ude

ABSTRACT

Nicotinate-phosphoribosiltransferase (NPRT) enzyme was isolated from Brevibacterium ATCC 6872 strain cells that carry out salvage way NAD overproduction. After 500-fold NPRT purification and its basic kinetic characteristics having been found (K_m for nicotinate, ATP and PRPP) the substance was defined as a monomeric protein with the molecular mass of 33.8 – 36.3 kD. The enzyme was found to be completely dependent on ATP ATP and slightly dependent on other nucleosidephosphates. It was discovered to be inhibited by the reaction products: nicotinic acid mononucleotide, pyrophosphate and ADP. Besides, NPPT, obtained from B.ammoniagenes, is subject to retroinhibition by the final product and by the NAD synthesis intermediate metabolite which is its desamidoNAD derivative. NADP does not affect the activity of NPRT produced from B.ammoniagenes ATCC 6872.

Key words: NAD, nicotinate-phosphoribosiltransferase, ATP, biosynthesis.

Nicotinamide adenine dinucleotide (NAD) is of great physiological significance for all living organisms. NAD and NADP participate as co-enzymes in hundreds of oxidative-reduction reactions. It is known that NAD (H) and NADP(H) can serve as allosteric effectors for the metabolism key enzymes. NAD is a component of ATP-ligaze reaction, it participates in ATP repairing processes and plays the role of a protector in detoxing cytotoxic compounds.

Besides supporting the de novo NAD and nucleoside phosphates synthesis cells are known to provide also salvage way of synthesis when the cells utilize exogenic precursors to

* East-Siberia State Technological University. Address: 670013 Ulan-Ude, Klyuchevskaya ul., 40A. E-mail: office@esstu.ru

synthesize pyridine and other types of nucleotides. It is this ability of cells that underlies the therapy process by using antimetabolites; it is also the basis for microbiological production of co-enzymes, nucleotides and their derivatives for the purpose of obtaining diagnostic preparations and medicines.

An important role in NAD salvage synthesis is played by phosphoribosiltransferase of nicotinic acid – an enzyme which is very labile to regulatory effects. This enzyme catalyses the reaction which results in the formation of nicotinic acid nucleotides from nicotinate and PRPP. It is the first stage in the salvage way NAD synthesis and it can limit the rate of the whole process. That is why studying this enzyme's regulation process is a part of the entire problem of cells functioning either at a normal or a pathological state and a part of NAD and its metabolites production and medical application (Bazdyreva, 1988).

The present paper studies the regulation process of nicotinate phosphoribosiltransferase. For this purpose NPRT-ase was purified and studies for its kinetic properties were undertaken – retroinhibiting effects of the reaction products, the effects of NAD synthesis metabolites on the enzyme activity.

Objects and Methods of Investigation

The objects of investigation are gram-positive bacteria Brevibacterium ammoniagenes ATCC 6872 which are peculiar for their remarkable activity in NAD and nucleoside phosphates salvage way synthesis. Nutrient media and cultivation conditions are described in (Tsirenov et al, 2004).

The quality of the biomass was determined spectrophotometrically by its optical density at 610 nm.

The quantity of the synthesized NAD was estimated by the fermentation method with alcoholdehydrogenase ((Tsirenov et al, 2004).

Cell-free extract was obtained by pressing cells with their successive extraction by 0.05 M–phosphate buffer (pH 7.4) and centrifugation at 20,000g for 20 minutes.

Column chromatography was performed by using DEAE-Sephacel columns (20×270mm and 14×190mm), DEAE – TOYOPEARL HW-65 (20×220mm) and Sephader G-150 (12×950mm).

The NPRT-ase activity was determined in the culture medium containing (mmol/ml): TRIS-phosphate buffer, pH 8.5-50; ATP.Na_2 -1; $MgCl_2$ – 10; PRPP.Mg_2-0.5, (^{14}C) nicotinic acid NA (50 Ku/M) – 0.1.

The total volume of the reactants was 250mkl. The reaction was initiated by adding 50 mkl of the enzyme preparation. Separating the reaction products was made on Silufol UV 254 plates by using a system of solvents: isobutiric acid: glacial acetic acid: 1N water ammonia (10:1:5)

The activity was evaluated by the transition of the radioactive trace from [^{14}C] – nicotinate to mononucleotide nicotinic acid (MNNA). Radioactivity rate was registered by means of the liquid scintillation counter SL – 4000 ("Intertechnique", France) .Enzyme activity unit was expressed as the amount of the reaction product being formed per minute. Mononucleotide of nicotinic acid concentration was determined spectrophotometrically taking into account the molar extinction index of MNNA (pH 7.0) by 275nm being equal to 43.10^3 (Honjo, 1971).

The enzyme's molecular mass was determined by gel-filtration in the Sephadex G-150 column (1000×12mm) and on 10% polyacrilamide gel plates (PAAG) with sodium dodecylsulphate present (SDS) (Martinussen et al, 2003).

Protein concentration was defined after (Lowry et al, 1951.)

RESULTS OF THE INVESTIGATION

Nicotinate-phosphoribosiltransferase catalyzes the first stage in the nicotinic branch of NAD synthesis.

Nicotinic acid (NA) + phosphoribosiltransferase (PRTP)+ATP → mononucleotide of nicotinic acid. (MNNA). At all stages nicotinate- phosphoribosiltransferase was isolated and purified at 4^0C. Purification results are given in Table 1.

Table 1. Nicotinate-phosphoribosiltransferase isolation from B.ammoniagenes ATCC 6872

	Purification Stage	Total protein, mg	Total activity, $E \cdot 10^{-3}$	Specific Activity, $E \cdot 10^{-3}$/mg	Activity yield, %	Degree of purification
1	Cell-free extract	3530	388.30	0.11	100	1.0
2	40-65% salting out $(NH_4)_2SO_4$	1200	324.00	0.27	84	2.5
3	Calcium-phosphate gel treatment	765	275.40	0.36	71	3.3
4	Acid treatment, pH 5.0	420	180.60	0.43	46	3.9
5	First DEAS -Sephacel chromatography	72	112,32	1,56	29	14,2
6	Second DEAS - Sephacel chromatography	17	94.35	5.55	24	50.5
7	DEAS – TOYOPEARL HW-65 chromotography	5	68.50	13.70	18	124.5
8	Sephadex G-150 gel-filtration	0.5	26.78	53.57	7	487.0

The molecular mass of the enzyme, determined by electrophoresis in PAAG in the presence of DDS-Na was equal to 33.8 kD and 36.6 kD by gel-filtration. According to the molecular mass of NPRT-ase from B.ammoniagenes ATCC 6872 to enzymes from other microorganisms and, like them, was a monomer.

Optimal conditions were selected to determine the enzyme activity.

Dependence of the reaction rate on its time was linear and equal to 30 min of incubation.

Research into the dependence of NAMN on reaction medium on synthesis of MNNA showed the enzyme to be active on a wide range of pH readings from 7.0 to 9.0, while optimal pH index was 8.5.

Dependence of the reaction rate on protein quantity is demonstrated in Fig. 1, which shows that the linear character of the dependence keeps in the range of 0.02 to 0.2 mg/ml during the 20 min incubation period.

The meaning of K_m for ATP determined according to Lineweaver-Berk coordinates was equal to $1.54 \cdot 10^{-4}$ M and did not depend on substrate concentration in the range $1 \cdot 10^{-5}$ - $1 \cdot 10^{-4}$ M for nicotinate acid NA and $1 \cdot 10^{-4}$ - $5 \cdot 10^{-4}$ M for PRPP. It is shown that other nucleosidetriphosphates, except ATP, can be used as energy-binding agents in MNNA-producing reaction, although less effectively. For example, the enzyme activity with CTP present was 80% of the activity in the presence of ATP. K_m for CTP was equal to $1.98 \cdot 10^{-4}$ M. Other nucleosidetriphosphates such as ITP, UTP and GTP irrespective of the base type, had less affinity to the enzyme and were less effective as cofactors.

Thus, the enzyme had greater affinity to ATP, but on the ATP concentration growing above 1 mm the substrate reaction inhibiting could be observed (K_i ATP was $9.5 \cdot 10^{-3}$M). It is shown that NPRT-ase is characterized by low specificity in relation to nucleosidetriphosphates, similar to its ATP dependent homologues obtained from other microorganisms.

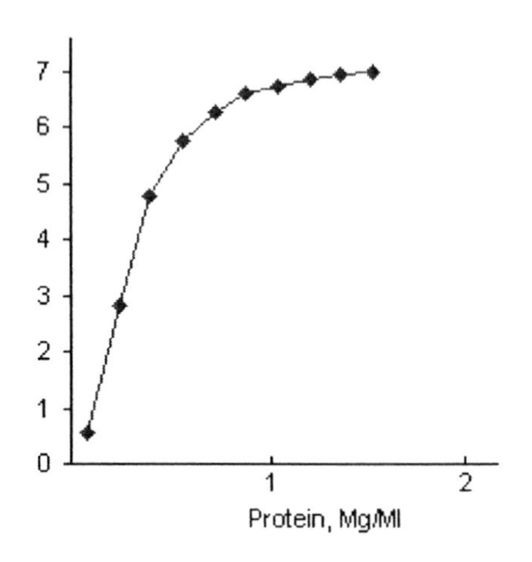

Fig. 1. Nicotinate - phosphoribosiltransferase activity as affected by protein concentration

The study of the reaction rate affected by NA concentration showed that K_m for nicotinate was equal to $0.6 \cdot 10^{-4}$ M (Fig. 2A, Table 2). Changes in PRPP concentration in the reaction medium from $6 \cdot 10^{-16}$ M to $6 \cdot 10^{-4}$ M and ATP concentration in the reaction medium from $5 \cdot 10^{-4}$ to $2 \cdot 10^{-3}$ M did not render any influence on the enzyme affinity to nicotinate.

Research into the reaction rate affected by PRPP concentration showed K_m for PRPP being equal to $1.45 \cdot 10^{-14}$ M (Fig. 2B, Table 2). Besides, it was shown that the double increase of nicotinic acid concentration from $5 \cdot 10^{-5}$ to $1 \cdot 10^{-4}$ M did not affect K_m for PRPP but resulted in 20% increase of the maximum reaction rate. PRPP concentration increasing above $6 \cdot 10^{-4}$ M resulted in enzyme inhibition by the substrate. K_i for PRPP obtained according to Dixon's coordinates, equaled to $1.7 \cdot 10^{-3}$ M, which was higher by an order of magnitude than K_m for PRPP (Tables 2, 3).

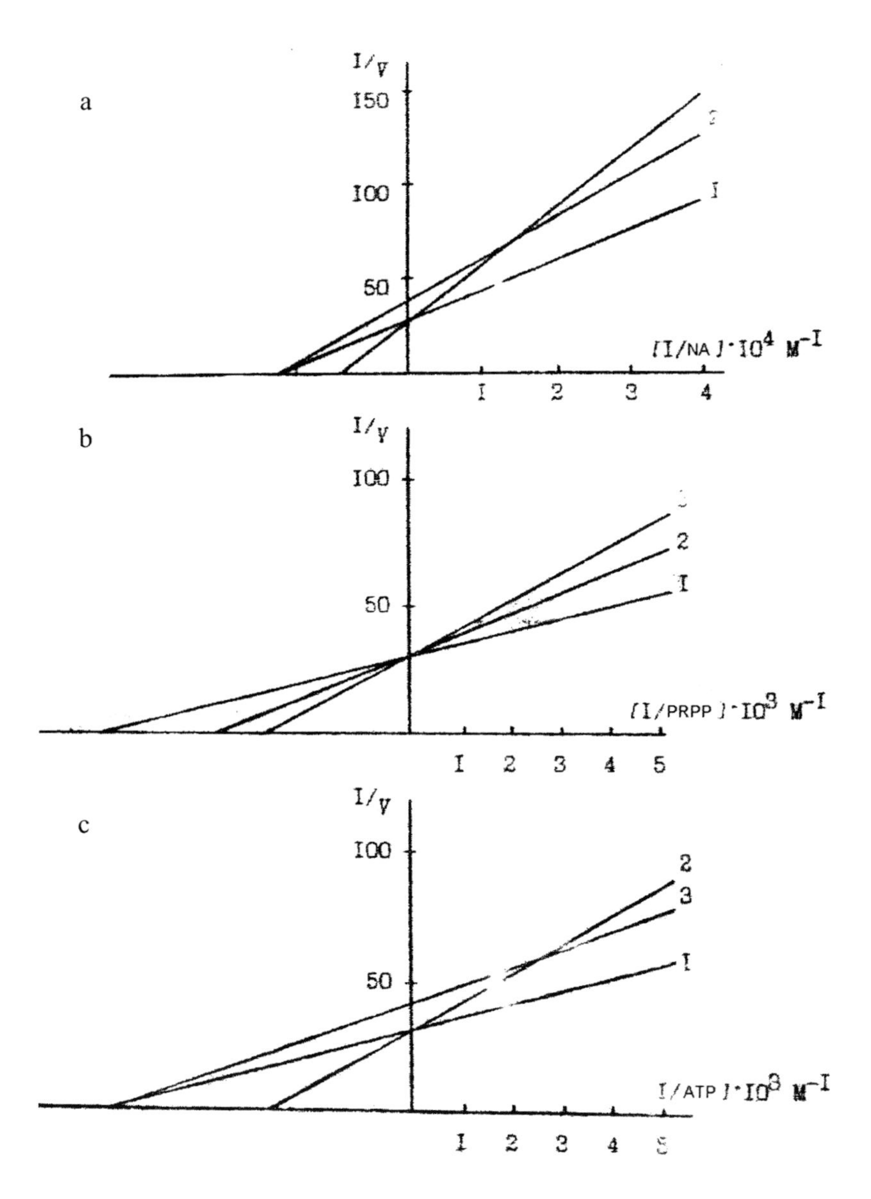

Fig. 2. Nicotinate - phosphoribosiltransferase from B. ammoniagenes activity as affected by reaction products
A – 1/V under NA concentration change;
B – 1/V under PRPP concentration change;
C – 1/V under ATP concentration change;
1 – control, 2 – pyrophosphate (0.6mm), 3 – MNNA (0.9mM).

The double increasing of nicotinic acid concentration (from $5 \cdot 10^{-5}$ to $1 \cdot 10^{-4}$M) did not affect K_m for PRPP but resulted in 20% maximum reaction rate growth. The increased concentration of PRPP (above $6 \cdot 10^{-4}$M) led to the enzyme inhibition by the substrate. K_i for

PRPP obtained according to Dixon's coordinates, was $1.7 \cdot 10^{-3}$ M which is higher by an order of magnitude than K_m for PRPP (Tables 2 and 3).

Purified NPRT-ase was used to investigate the enzyme activity affected by end reaction products. MNNA was shown to inhibit NPRT-ase (K_i $1.9 \cdot 10^{-3}$M). In this case the inhibition constant was equal to $1.9 \cdot 10^{-3}$M. NAMN demonstrated concurrent inhibition in relation to nicotinic acid and PRPP, reducing the enzyme affinity to the substrates (Table 2). In respect to ATP MNNA inhibition type was non-concurrent (Table 3). Niacin derivatives, nicotinamide and nicotinamide mononucleotide did not affect the enzyme activity.

Table 2. NAD synthesis metabolites affecting the affinity of Nicotinate-phosphoribosiltransferase substrates obtained from B.ammoniagenes ($K_m \cdot 10^{-4}$ M)

	Substrates		
	NA	PRPP	ATP
Control	0.60	1.45	1.54
NAMN	1.0	3.22	1.54
PP	0.60	2.52	3.13
DesamidoNAD	0.60	1.45	1.54
NAD	0.60	1.45	2.50

Table 3. Activity of Nicotinate-phosphoribosiltransferase obtained from B. ammoniagenes inhibited by NAD ($K_i \cdot 10^{-3}$M) salvage synthesis metabolites

ATP	9.5	MNNA	1.9
ADP	1.2	desamidoNAD	1.1
PRPP	1.7	NAD	0.7
PP	0.9	-	-

One of the final products – pyrophosphate (PP) demonstrated 40% inhibition of the enzyme activity. K_i for PP was equal to $9 \cdot 10^{-4}$M. Pyrophosphate was a concurrent inhibitor in relation to ATP and PRPP.

ADP behaved as a concurrent inhibitor in relation to ATP with K_i equal to $1.2 \cdot 10^{-3}$M. Nucleosidephosphates GDP, IDP, UDP, CDP and nucleosidemonophosphates AMP, GMP, IMP, CNP and TMP of 1mM concentration did not produce any effect on NPRT-ase.

Adenine, adenosine, adenosine – 5 – teraphosphate and polyadenin acid did not affect NPRT-ase activity.

NPRT-ase was found to be the object of NAD inhibiting its own synthesis. NAD was found to produce stronger inhibition as compared to other niacin metabolites (Table 3).In the concentration of 0.5 mM it produced about 30% reduction of the enzyme activity. K_i equaled to $0.7 \cdot 10^{-3}$M. NAD behaves as a concurrent/non-concurrent inhibitor in relation to ATP, increasing Km up to $2.50 \cdot 10^{-4}$M.

DesamidoNAD affected the enzyme activity (K_i, 1.1·10-3M) less. Inhibition was non-concurrent in relation to NA, PRPP and ATP.

Thus, NPRT-ase is the second site (after adenilatekinase) preparation affected by NAD and its desaminated derivative desamidoNAD retro-inhibiting behavior.

Analysis of data published shows that the activity of NPRT-ase, isolated from animal tissues and a number of microorganisms, has not been inhibited by desamidoNAD, NAD and NADP until now.

The strictly limited content of pyridine co-enzymes characteristic of a cell in its normal development is probably supported by other enzymes of this biosynthetic way regulation.

So, various ways of the enzyme activity control, which involves almost all metabolites of NAD synthesis salvage way, confirms the NPRT-ase key role in the pyridine nucleotides supersynthesis for the strain in question.

Conclusion

1. The nicotinade-phosphoribosiltransferase (NPRT-ase) enzyme was isolated from B.ammoniagenes ATCC 6872 cells and subjected to 500-fold purification.
2. NPRT-ase basic kinetic characteristics were determined and it was shown that NPRT-ase was a monomeric protein, its molecular mass being 33.8 – 36.3 kD.
3. Complete dependence of NPRT-ase on ATP and the enzyme's low nucleotide specificity were proved.
4. The type of NPRT-ase inhibition by the reaction products MNNA, pyrophosphate and ADP was found.
5. NPRT-ase retroinhibition by the end product and the intermediate metabolite of NAD synthesis – its derivative desamidoNAD – was demonstrated, while NADP did not affect NPRT-ase activity.

References

[1] N.M. Bazdyreva: *Appl. Biochemistry and Microbiology*, V. XXIV, W2 147 – 163 (1988 (In Russia)

[2] V.ZH. Tsirenov, L.V. Tulohonova, E.M. Podlepa et al. In: *Biotechnology and Industry*. Editor G.E. Zaikov, pp.55 – 74, (2004, Nova Science Publishers, Inc.)

[3] G.A. Kochetov

[4] Honjo, Y. Nishizuka *Meth. Enzymol*, v. 18, pp 132 – 137, (1971)

[5] J. Martinussen, S.L. Wadskov-Hansen and K. Hammer *Bacteriology,* vol. 185, N4. pp. 15003 – 15008 (2003)

INDEX

H

I

J

K

L

M

N

O

P

T

U

V

W

Y